Springer Undergraduate Mathematics Series

SUMS Readings

Editor-in-Chief

Endre Süli, Oxford, UK

Series Editors

Mark A. J. Chaplain, St Andrews, UK

Angus Macintyre, Edinburgh, UK

Shahn Majid, London, UK

Nicole Snashall, Leicester, UK

Michael R. Tehranchi, Cambridge, UK

SUMS Readings is a collection of books that provides students with opportunities to deepen understanding and broaden horizons. Aimed mainly at undergraduates, the series is intended for books that do not fit the classical textbook format, from leisurely-yet-rigorous introductions to topics of wide interest, to presentations of specialised topics that are not commonly taught. Its books may be read in parallel with undergraduate studies, as supplementary reading for specific courses, background reading for undergraduate projects, or out of sheer intellectual curiosity. The emphasis of the series is on novelty, accessibility and clarity of exposition, as well as self-study with easy-to-follow examples and solved exercises.

Stefan Steinerberger

The Unreasonable Elegance of Mathematics

Stefan Steinerberger iD
Department of Mathematics
University of Washington
Seattle, WA, USA

ISSN 1615-2085   ISSN 2197-4144 (electronic)
Springer Undergraduate Mathematics Series
ISSN 2730-5813   ISSN 2730-5821 (electronic)
SUMS Readings
ISBN 978-3-032-03814-2   ISBN 978-3-032-03815-9 (eBook)
https://doi.org/10.1007/978-3-032-03815-9

Mathematics Subject Classification: 00-01, 00A09, 00A06

Anaximenes to Pythagoras

Thales, the son of Euxamias, has died in his old age, by an unfortunate accident. In the evening, as he was accustomed to do, he went forth out of the vestibule of his house with his maid-servant, to observe the stars: and (for he had forgotten the existence of the place) while he was looking up towards the skies, he fell down a precipitous place. So now, the astronomer of Miletus has met with this end. But we who were his pupils cherish the recollection of the man, and so do our children and our own pupils: and we will lecture on his principles. At all events, the beginning of all wisdom ought to be attributed to Thales.

Diogenes Laertius, The Lives and Opinions of Eminent Philosophers, 3rd century AD

Preface

Origin. The origin of this book can be traced to a very specific moment. During a faculty meeting it was pointed out that the Department of Mathematics at the University of Washington does not offer an introductory math course for people who are *not* math majors. Students with an interest in political science or biology or architecture had no chance to find out what math actually was: sometimes there was a basic course (usually Calculus) that they *had* to take. While History offers 'Introduction to History' and Biology offers 'Biology 101', there was no such course in Mathematics. *What's the difficulty?*, I remember asking, *Creating such a course should be fairly straightforward.* Little did I know that I had just committed a basic mistake, stating an opinion. As best said by the French mathematician Blaise Pascal, *I have discovered that all the unhappiness of men arises from one single fact, that they cannot stay quietly in their own chamber* [223]. Indeed, it was immediately agreed that we should create such a course, that I should be in charge of creating it, and that I should teach the first instance (and, as it would turn out, the second).

Math 180. The course was given the provisional name 'Math 180' and it was my job to come up with a name. Being short on ideas, I decided to give it the worst name I could think of. Surely someone would come and protest and suggest a better name. Nobody came, nobody protested, and suddenly I found myself teaching *A Walk in the Garden of Mathematics* in the Winter Quarter 2024. The initial run had 12 heroic students from all over the University, including students of Biology, students of Architecture, and students of English Literature. I was unable to find a book that captured what I was trying to do and decided to write some lecture notes. Habiba Kalantarova liked the lectures notes and here we are.

What this book might be good for. Books are magical. Maybe this one is just the right size to even out your crooked table. I have no idea what it might end up getting used for, but I can list some possible applications that I had in mind.

1. As a general text to be enjoyed by people who like mathematics and would like to learn more about it and its history. The history of mathematics is far from straightforward. Its protagonists are alchemists and wizards, half of them busy

with predicting the apocalypse. It is an enormously enjoyable vantage point from which to view humanity, its moments of glory and folly.[1]

2. As a survey of unusual ideas and even more unusual characters.[2] It is a pity that so few people have ever heard of Roger Bacon or John Duns Scotus. If even one reader is induced to read Diogenes Laertius' *The Lives and Opinions of the Eminent Philosophers*, well, that would be something. Boswell's *Life of Samuel Johnson* has entertained people for more than 200 years, it may entertain you. Cardano's *The Book of My Life* is a real page turner. Maybe some of the characters showing up in this book can suggest your next book.

3. As a basis for general introductory lectures for non-math majors. With such an audience in mind, I have tried to keep the technical level firmly below the calculus level. Large portions can be enjoyed with just basic high school algebra.

4. When I became interested in mathematics, my parents would buy me introductory books, 'What is mathematics?'. These books contributed to my interest in the subject (more than some teachers). If this text can serve a similar purpose for even one person, I would consider it a resounding success.

5. My parents never stopped buying me introductory math books: long after becoming a professor, they would still reliably give me some version of 'An Introduction to Mathematics' as a present (I got one just 6 weeks ago). Maybe this text can serve as a gift for someone who is either already interested in mathematics or interested in learning more about why *some* people seem to love it so much. Spouses of mathematicians who end up at a table filled with mathematicians seem to experience unusual levels of agony. Maybe this book can help. And maybe my parents will stop buying me such books after I finally write one myself.

Thanks. I am grateful to many people. Remi Lodh was a wonderful editor. François Clément, Ryan North, Andrea Ottolini, Erich Steinerberger helped in different ways. I am indebted to the University of Washington Librarians with a special thanks to Maryam Fakouri. I am indebted to Patrick Hanslmaier for a thoughtful and very extensive critical reading. He and Mariana Smit Vega Garcia had to listen to an endless stream of math history for about a year, sorry about that.

Stefan Steinerberger
Seattle, June 2025

---

[1] With some emphasis on the folly; there is often too much focus on the brilliant insight and not enough on the flawed human who had it. Yes, Pythagoras was probably a genius but that does not mean that he was not also weirdly obsessed with beans.

[2] Most characters in the book are European men. This is partially explained by the fact that women were not allowed to participate in the scientific enterprise for most of history. The geographic focus is a consequence of my own limitations: I wanted to engage with the source material as much as possible and do not speak Arabic, Chinese or Sanskrit. I regret that the reader has to miss out on the House of Wisdom, Ibn al-Haytham, Bhāskara II, Yang Hui and many others.

Contents

Chapter 1
What is Mathematics?

What exactly is mathematics? Many have tried but nobody has really succeeded in defining mathematics; it is always something else. (Stanislaw Ulam, [286])

1.1 So, what is it?

The very first question is perhaps how to define *Mathematics*. If one defines geology as the study of rocks and linguistics as the study of language, one may be inclined to define mathematics as the study of numbers. All three definitions are incorrect or, at the very least, incomplete. They miss the point. Luckily, a geologist does not need to be able to define geology to discover new structures in a volcano (this proves that the author does not know what geologists actually do); a linguist does not need to have a precise definition of 'language' to study local dialects. Similarly, we will be perfectly able to have fun with mathematics without ever precisely defining what it is. Indeed, no general definition exists; if one asks five mathematicians how they define their work, one is likely to receive five very different replies. However, it may be helpful to have at least some approximation: as a starting point, we define

Mathematics is the study of formal structures.

Here, 'formal' means that statements, arguments and processes can be made so very precise that a computer can verify each step. Examples of things that would belong to mathematics in this sense would be

1. a computation, $5 + 8 = 13$
2. a method that can be used to discover whether an integer is a prime number (only divisible by 1 and itself: 2, 3, 5, 7, 11, ...) or not
3. a rigorous argument showing that in a right-angled triangle $a^2 + b^2 = c^2$
4. a computation on the Poker channel showing that one of the players has a hand that has a 73% chance of winning
5. an argument showing that, in a specific game of chess, one of the two players can force a win if they play flawlessly.

These are very different structures. One example concerns the rules of addition of integers $1, 2, 3, \ldots$, another multiplication. The third example, Pythagoras' Theorem, is a statement in a setting where the rules are geometric: lengths, angles, distances. The fourth example is a statement about probabilities (and what exactly it means to be a probability is interesting in itself, see §16). The last example is set within the formal system of chess: it may be very different from, say, the integers or triangles, but it is a completely well-defined formal structure.

© The Author(s), under exclusive license to Springer Nature Switzerland AG 2025
S. Steinerberger, *The Unreasonable Elegance of Mathematics*,
Springer Undergraduate Mathematics Series, https://doi.org/10.1007/978-3-032-03815-9_1

1.2 Formal systems

One thing that all these examples have in common is that there is no room for debate: if we were to take two robots and explain the rules of chess to them and then ask them whether, in a specific instance, one of the players can guarantee a win, the two robots have to come to the same conclusion. Either there exists a way a player can move the pieces to force a win (meaning that, independently of what the other player does, there is always a way to react that leads to victory) or there does not. Chess is a finite game, the answer always exist. Whether humans, using computers at our current state of technology, can find the answer, that is a different question: the answer exists. A more colorful definition would be

> Mathematics is the study of structures where two reasonable people can never disagree.

If we are given a formal system and all its rules and we are asked a specific question (can 18 be written as the product of two smaller integers?), then there always exists an answer. It may well be that we are unable to find the answer quickly (can 23487235081 be written as the product of two smaller integers?) and maybe we do not know how to find the answer at all (or things may be even weirder than that, see §19). However, if one person proposes an answer (yes, 18 can be written as the product of two smaller integers) and shows their reasoning within the formal system (3 and 6 are smaller than 18 and $3 \cdot 6 = 18$), then there is no way of disagreeing with this. Feel free to disagree with this definition of Mathematics or to come up with your own. Luckily, we do not need to define mathematics to enjoy it.

There are infinitely many formal systems. For example, one could play chess on an 8×8 board or on a 9×9 board (maybe with some new pieces added) or on a 10×10 board and so on. All of these would constitute valid formal systems in which one could ask questions. But do you really care about a particular chess configuration on a 615×615 chess board? Probably not. Mathematics studied by people is usually

1. beautiful (we'll discuss many examples; beauty is subjective, naturally)
2. useful to other sciences (happens frequently)
3. or both (also happens very frequently).

Not all formal structures are like that – but if it is not beautiful and it is not useful, why should we care about it? Most people are not very interested in the aesthetic aspects of the complete theory of chess on a 615×615 chess board; it is also not going to help you much in daily life. One of the most fascinating things about mathematics is that there are things that we do **not** know. It is one of the most widely spread misconceptions that mathematics is essentially finished ('Don't you already know all the numbers?'). Mathematics is sometimes taught as a sequence of seemingly arbitrary rules that one has to memorize – not too different from the way there are rules in chess. Math is more than that: we discover new things every day, not necessarily new rules but new consequences of the rules. Much like geology or linguistics, mathematics is an evolving science! Scientists like to think about things that are not yet understood, that's where all the fun is. Mathematicians are very similar: they like to think about things that we do not yet understand.

1.3 Aesthetic qualities

Mathematics is a powerful tool and can be used to describe reality: but that is not the only reason why people are interested in it. Much of the interest is motivated by the fact that it is extraordinarily beautiful.

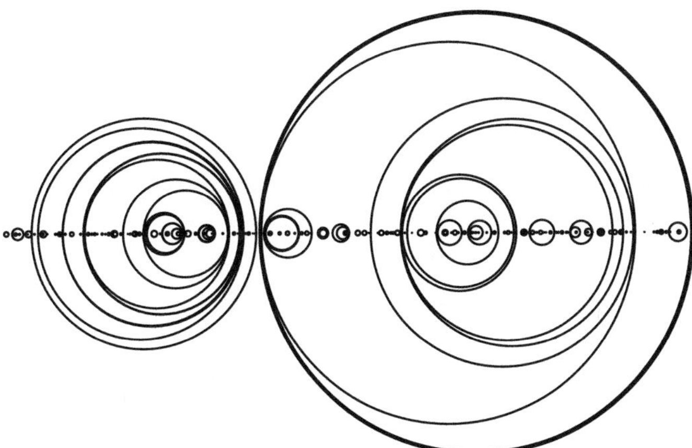

Fig. 1.1: A very pretty arrangement of circles arising from studying which pizzeria delivers pizza to which house (see §17 for more details).

We quickly show some pictures that arose organically in the author's work (as opposed to: the standard propaganda pictures handed out to mathematicians when they get their PhD and are told to go forth and proselytize). The first picture, Fig. 1.1, shows the solution of a logistic (delivery) problem: visualizing the optimal solution leads to some beautifully arranged circles.

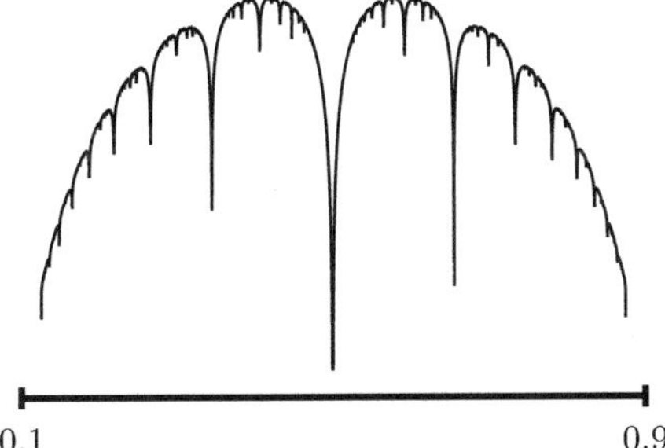

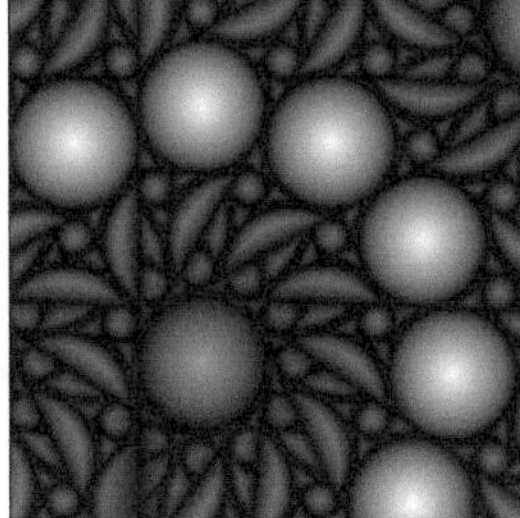

Fig. 1.2: Left: a function with some very interesting behavior [268]. Right: a picture arising from the study of a very simple way of adding and subtracting numbers [271].

More examples are shown in Fig. 1.2. On the left, we see the graph of the function

$$f(x) = |\sin(x)| + |\sin(2x)|/2 + |\sin(3x)|/3 + \cdots + |\sin(nx)|/n,$$

where $n = 50000$ (and the picture gets more and more interesting as n increases). For n very large, we know that the function has a local minimum at all rational numbers whose denominator is 'small' compared to n. This in itself is curious, however, there are many other things about the function that remain unknown. The image on the right, well, we do not understand it at all, see [271] for more details.

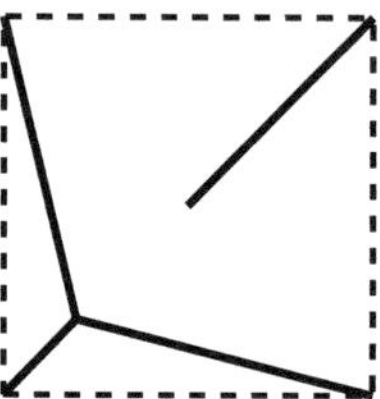

Fig. 1.3: Shortest known opaque set with length $\sqrt{2} + \sqrt{3/2} \sim 2.63$.

We conclude with an example of something that we do not yet understand [152, 269]. The set of four line segments shown in Fig. 1.3 has the property that every line that goes through the square is also going to hit that set. Such sets are called *opaque*. If we take all 4 sides of the square, the boundary of the square, then this set is opaque: any line hitting the square will also hit the boundary. This is an opaque set with length 4. We could even omit one of the four sides, taking three sides would be enough, that would be an opaque set with length 3. What is the length of the shortest opaque set? It is probably the one in Fig. 1.3, but we do not know!

1.4 Exercises

1. Come up with a formal definition of an everyday object (say, a sandwich) and have a friend try to find borderline cases (is a Taco a sandwich?).
2. Learn about Chess variants, some of these are truly amazing (Fischer Random Chess, 3D Chess, Dark Chess, ...)
3. Consider the problem of finding the shortest opaque set for the unit disk. Taking the boundary of the disk would give an opaque set with length 2π. Can you construct a shorter one?

Chapter 2
Pascal's Triangle, Number Magic and the Apocalypse

The last thing one settles in writing a book is what one should put in first. (Pascal [223])

2.1 Pascal's Triangle

We start with an old construction: Pascal's triangle. It is a formal system based on adding integers and named after the French philosopher and mathematician Blaise Pascal (1623–1662). However, the idea is *much* older than Pascal and goes back at least to the Persian mathematician Al-Karaji (953–1029). This leads to a first natural question: why do we call it Pascal's Triangle and not Al-Karaji's triangle? This is a common theme in mathematics. The Russian mathematician Vladimir Arnold noted

> Similarly to the fact that America does not carry Columbus's name, mathematical results are almost never called by the names of their discoverers. [...] Prof. M. Berry once formulated the following two principles
> *The Arnold principle.* If a notion bears a personal name, then this name is not the name of the discoverer.
> *The Berry principle.* The Arnold Principle is applicable to itself. (V. Arnold [13])

Berry is right: the statistician Stephen Stigler published what he called *Stigler's Law of Eponymy*, summarizing it as *No scientific discovery is named after its original discoverer* and ascribing it to Robert Merton: *If there is an idea in this paper that is not at least implicit in Merton's* THE SOCIOLOGY OF SCIENCE, *it is either a happy accident or a likely error* [274]. There are many examples: the Pythagorean Theorem predates Pythagoras by at least a thousand years. Keeping this in mind, we can introduce the Pascal Triangle as an infinite triangle of numbers. When writing down a new row, first put a 1 on each side and, in the middle, make each number the sum of the two numbers above.

$$
\begin{array}{ccccccccccccc}
 & & & & & & 1 & & & & & & \\
 & & & & & 1 & & 1 & & & & & \\
 & & & & 1 & & 2 & & 1 & & & & \\
 & & & 1 & & 3 & & 3 & & 1 & & & \\
 & & 1 & & 4 & & 6 & & 4 & & 1 & & \\
 & 1 & & 5 & & 10 & & 10 & & 5 & & 1 & \\
1 & & 6 & & 15 & & 20 & & 15 & & 6 & & 1
\end{array}
$$

Fig. 2.1: The first few rows of Pascal's triangle.

© The Author(s), under exclusive license to Springer Nature Switzerland AG 2025
S. Steinerberger, *The Unreasonable Elegance of Mathematics*,
Springer Undergraduate Mathematics Series, https://doi.org/10.1007/978-3-032-03815-9_2

One can think of it almost as a game, we start with 1 and then keep adding previous numbers to produce this triangular shape. This triangular arrangement of numbers can now be investigated. Pascal's Triangle has so many nice properties that one could spend *years* studying it (and many people have). Perhaps the most fundamental property is that it shows up naturally when multiplying out expressions. Pascal's triangle describes the numbers showing up in the Binomial Theorem, which generalizes the familiar $(a+b)^2 = a^2 + 2ab + b^2$,

$$(a+b)^0 = \mathbf{1}$$
$$(a+b)^1 = \mathbf{1} \cdot a + \mathbf{1} \cdot b$$
$$(a+b)^2 = \mathbf{1} \cdot a^2 + \mathbf{2} \cdot ab + \mathbf{1} \cdot b^2$$
$$(a+b)^3 = \mathbf{1} \cdot a^3 + \mathbf{3} \cdot a^2b + \mathbf{3} \cdot ab^2 + \mathbf{1} \cdot b^3$$
$$(a+b)^4 = \mathbf{1} \cdot a^4 + \mathbf{4} \cdot a^3b + \mathbf{6} \cdot a^2b^2 + \mathbf{4} \cdot ab^3 + \mathbf{1} \cdot b^4$$

2.2 A Number Magician: Michael Stifel

Pascal's triangle was studied by very interesting people. A good example is its appearance in *Arithmetica Integra*, a book written by Michael Stifel (1487–1567) in 1544 and published 79 years before Pascal was even born.

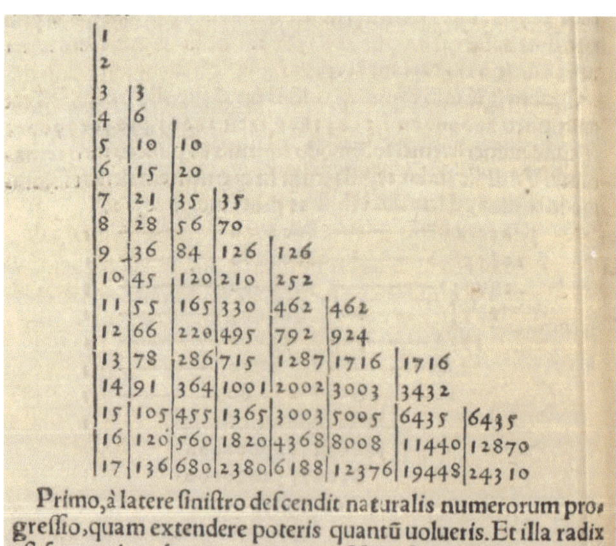

1							
2							
3	3						
4	6						
5	10	10					
6	15	20					
7	21	35	35				
8	28	56	70				
9	36	84	126	126			
10	45	120	210	252			
11	55	165	330	462	462		
12	66	220	495	792	924		
13	78	286	715	1287	1716	1716	
14	91	364	1001	2002	3003	3432	
15	105	455	1365	3003	5005	6435	6435
16	120	560	1820	4368	8008	11440	12870
17	136	680	2380	6188	12376	19448	24310

Primo, à latere finiftro defcendit naturalis numerorum pro-
greffio, quam extendere poteris quantū uolueris. Et illa radix

Fig. 2.2: Pascal's Triangle as it appears in Michael Stifel's *Arithmetica Integra* [273].

Medieval mathematicians were often also magicians, alchemists or prophets. Michael Stifel is no exception. Ordained as a priest in the Order of Saint Augustine, Stifel got interested in mathematics because he was doing 'number magic'. In 1532, before his more serious mathematical studies, he published (anonymously) *A little calculation book about the Antichrist*. This is a truly remarkable book because it illustrates an unusual ('crazy') way of reasoning and is somewhat representative for certain branches of numerology. Stifel first proposes (see Fig. 2.3) how one might naturally associate a number to each letter ($a = 1, b = 2, \ldots$). The 'value' of a word is then sum of the numbers associated to each letter. Stifel illustrates this with the word 'Hvsz', presumably referring to the Czech theologian Jan Hus (1370–1415). He writes *put for the H 8, for the v 20, for the s 18 and for the z 23. So maketh the entire word 69.* He is also very peculiar about the letters he is using and writes *The w is not needed in Latin, hence one should not use it.*

1	2	3	4	5
a	b	c	d	e
6	7	8	9	10
f	g	h	i	k
11	12	13	14	15
l	m	n	o	p
16	17	18	19	20
q	r	s	t	v
21	22	23.		
x	y	z.		

1	3	6	10	15
a	b	c	d	e
21	28	36	45	55
f	g	h	i	k
66	78	91	105	120
l	m	n	o	p
136	153	171	190	210
q	r	s	t	v
231	253	276		
x	y	z		

Fig. 2.3: Stifel's *Rechenbuchlin* (1532) explaining two different system of how to associate letters with numbers. One uses integers, the other triangular numbers.

Science and progress are rarely ever linear, frequently there are leaps of intuition and sometimes these leaps lead absolutely nowhere. Stifel illustrates this nicely.

> After I had fun for a while with these types of computations and discovered some results that surprised me I started to wonder whether one may calculate, in this manner, the number of the Beast in the Apocalypse. But I did not find a name that would reach the number of the beast, that is 666. So it occurred to me to increase the numbers .... (Michael Stifel [272])

Stifel proceeds to propose a different way to associate numbers to letters (see Fig. 2.3, right). The numbers appearing in his second proposal, $1, 3, 6, 10, 15, 21, 28, 36 \ldots$, are the sums of the first few integers $1, 1 + 2 = 3, 1 + 2 + 3 = 6, \ldots$ and so on. They are, in particular, *the second line in the Pascal triangle*. Equipped with his new system, Stifel starts doing 'mathematics' and investigates the name MARTIN LUTHER, who was born in 1483. He computes

> And so I put down the name MARTINUS LUTER [*sum 1573*] and found it to be bigger than his year of birth and so removed the 'n'. This resulted in the name MARTIUS LUTER [*sum 1482*] and Martius means like a warrior [*Mars is the god of war*] which I like [...] because of his war against the Pope [...] but now through omission of the letter 'n' the year is 1 too small [...] but if I write instead LAUTER [*german: 'honest'*] then the year of his birth fits. Therefore I write MARTIUS LAUTER and this leads to 1483 [...] Observe what kind of mysteria are contained in this name. (Michael Stifel [272])

This is certainly a 'curious' way to reason. Stifel, having published his Antichrist computations in 1532 promptly predicted the end of the world (*Apocalyps in Apocalypsim*) in the year 1533 (on October 19 at 8am). When the world continued to exist, Stifel turned towards more serious mathematics. In 1541, at the age of 54, he enrolled at the University of Wittenberg to deepen his understanding of mathematics and in 1558, at the age of 71, he is listed as affiliated with the University of Jena. He is now remembered for introducing the notion of powers such as $2^3 = 8$ but also, importantly, $2^{-3} = 1/8$ so that $2^a \cdot 2^b = 2^{a+b}$ is true for all integers a, b (not only positive ones). The idea of multiplying a number by itself, like $2^3 = 2 \cdot 2 \cdot 2 = 8$, is quite natural. The idea of doing this a negative number of times, multiplying a number -3 times by itself, that is radically new, like many of Stifel's ideas.

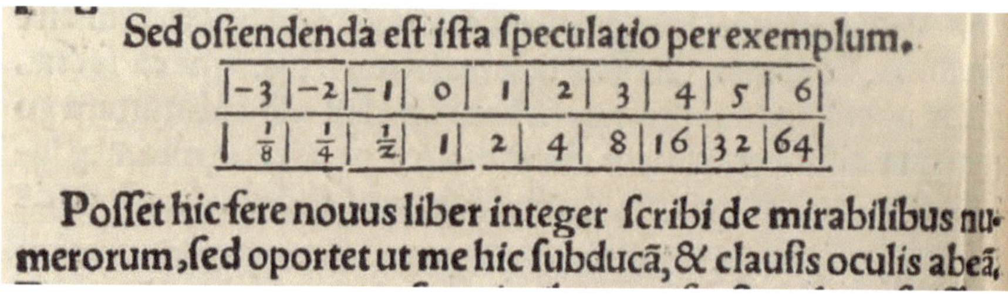

Fig. 2.4: $2^{-3} = \frac{1}{8}$ and $2^3 = 8$ in *Arithmetica Integra* [273]. The text is heartbreaking: *But this theory must be demonstrated with an example. [Table] An entire new book could be written here about the wonders of the numbers, but I must withdraw here and leave with my eyes closed.*

2.3 Factorials

Leaving number magic behind and returning to the Pascal triangle, is it possible to predict the entries of the triangle without computing them from scratch? Surely, if I want to know the 19th entry of the 1533th row, I could start writing down the triangle but it would take a long time. There is a more direct way using *factorials*. For any positive integer n, we use the expression $n!$ (spoken 'n factorial') to refer to

$$n! = 1 \cdot 2 \cdot 3 \cdot 4 \cdots n$$

which is the multiplication of the first n integers. For example $3! = 1 \cdot 2 \cdot 3 = 6$ and $5! = 1 \cdot 2 \cdot 3 \cdot 4 \cdot 5 = 120$. The sequence of factorials begins $1, 2, 6, 24, 120, 720, 5040, 40320, \ldots$ and people have thought about these numbers for a very long time. They show up quite naturally in one of the oldest books of Jewish mysticism, the *Sefer Yetzirah*. As is appropriate for a book of mysticism, its origin remains mysterious and even its age is unknown (possibly some time between the second and sixth century).

> Two stones build two houses. Three stones build six houses. Four stones build twenty-four houses. Five stones build one hundred twenty houses. Six stones build seven hundred twenty houses. Seven stones build five thousand forty houses. Thenceforth, go out and calculate what the mouth is unable to say and what the ear is unable to hear. (*Sefer Yetsirah*)

These numbers are just $2!, 3!, 4!, 5!, \ldots$, the factorials! Likewise, Plato, when discussing the ideal size of a city, comes to an interesting conclusion: he argues that the ideal city has $7! = 1 \cdot 2 \cdot 3 \cdot 4 \cdot 5 \cdot 6 \cdot 7 = 5040$ inhabitants. This number, Plato argues, is good because it can be evenly divided into groups of many different sizes.

> Every legislator ought to know so much arithmetic as to be able to tell what number is most likely to be useful to all cities; and we are going to take that number which contains the greatest and most regular and unbroken series of divisions. The whole of number has every possible division, and the number 5040 can be divided by exactly fifty-nine divisors, and ten of these proceed without interval from one to ten: this will furnish numbers for war and peace, and for all contracts and dealings, including taxes and divisions of the land. [232]

Plato does not explain what he plans to do about people giving birth ($5041 = 71^2$ can only be divided by 71) or, even worse, people dying (5039 is a prime number and is not divisible by any number). The factorial numbers are already filled with mysteries and there are many things we do not know about them. Plato's example shows that $7! = 5040 = 71^2 - 1$ and we could look for other factorial numbers that are one smaller than a square number. More formally, we are looking for solutions of the equation $n! + 1 = m^2$ where n and m are some integers. Playing with some examples, we find

$$4! + 1 = 25 = 5^2 \qquad 5! + 1 = 121 = 11^2 \qquad 7! + 1 = 71^2.$$

It is believed that these are the only solutions and that no other solutions exist. This looks like a pretty simple question, it does not involve much more than multiplication, but we still do not know – and this question is more than a 100 years old.

Fig. 2.5: Brocard [38] asking in 1885 his *Question 1532. For which values of x is the expression* $1.2.3.4\ldots.x + 1$ *a perfect square?*

The belief that there are no other solutions except the ones listed above is known as the Brocard–Ramanujan conjecture ('conjecture' being a way of saying 'things we believe to be true but we do not know for sure'). Among the many properties of factorials, a curious one is that they know what prime numbers are. It starts with an innocent question: when is $(n-1)! + 1$ divisible by n? Looking at the examples, we

n	2	3	4	5	6	7	8	9	10	11
$(n-1)! + 1$	2	3	7	25	121	721	5041	40321	362881	3628801
divisible by n	✓	✓	×	✓	×	✓	×	×	×	✓

see that the values of n for which this is true are given by $2, 3, 5, 7, 11, \ldots$ It appears to be the sequence of prime numbers, numbers that are only divisible by 1 and themselves. This magical property of the factorials is known as *Wilson's Theorem*, first announced by Edward Waring.

$$\text{Sit } n \text{ numerus primus, \& } \frac{1 \times 2 \times 3 \times 4 \ldots \overline{n-2} \times \overline{n-1} + 1}{n} \text{ erit}$$

$$\text{integer numerus, e. g. } \frac{1 \times 2 + 1}{3} = 1, \quad \frac{1 \cdot 2 \cdot 3 \cdot 4 + 1}{5} = 5,$$

$$\frac{1 \cdot 2 \cdot 3 \cdot 4 \cdot 5 \cdot 6 + 1}{7} = 103, \text{ \&c.} \quad \text{Hanc maxime elegantem pri-}$$

morum numerorum proprietatem invenit vir clariffimus, rerum que mathematicarum peritiffimus Joannes Wilfon Armiger.

Fig. 2.6: Waring's 1770 announcement [294] of Wilson's Theorem.

If n is a prime number, & $\frac{1 \times 2 \times 3 \times 4 \ldots (n-2) \times (n-1) + 1}{n}$ is a whole number, e.g. $\frac{1 \times 2 + 1}{3} = 1$, $\frac{1 \cdot 2 \cdot 3 \cdot 4 + 1}{4} = 5$, $\frac{1 \cdot 2 \cdot 3 \cdot 4 \cdot 5 \cdot 6 + 1}{7} = 103$, &c. This most elegant property of prime numbers was discovered by a most famous man, and expert in mathematical things, John Wilson, Esquire. (Edward Waring [294])

Waring's 1770 announcement did not have a proof. A year later, the Italian mathematician Joseph-Louis Lagrange called this *a very beautiful statement of the arithmetic* and gave a proof [162]. Factorials grow very quickly and it is relatively easy to see (see Exercises) that

$$2^{n-1} \leq 1 \cdot 2 \cdot 3 \cdots (n-1) \cdot n \leq n^n.$$

A much more precise and also much more surprising estimate is known as the *Stirling approximation*, named after James Stirling (1692–1770). Stirling had an exciting life: in 1717 he moved to Venice (leading to him acquiring the wonderful nickname *The Venetian*). There, he got into trouble because of what one would now call industrial espionage. Indeed, in Venice *he had, at the request of certain London merchants, acquired information regarding the manufacture of plate glass. [...] owing to his discovery he had to flee from Venice, his life being in danger* [284]. Stirling's approximation is equally exciting and deals with the question of how large we expect $n!$ to be. Stirling's approximation is

$$n! \sim \sqrt{2\pi n} \left(\frac{n}{e}\right)^n,$$

where $\sim$ means that the ratio of the two quantities approaches 1 when n becomes very large. This formula is *shocking*: on the left-hand side, we have the product of the first n integers (that is $n!$) and on the right-hand side we have a formula involving both π, the area of the unit disk, and Euler's number e. How do the integers know about π, how do they know the area of the disk with radius 1?

2.4 Binomial Coefficients

Returning to Pascal's triangle, we can, using factorials, give a short and simple formula for its entries.

$$\text{The } k\text{-th entry in the } n\text{-th row} \quad = \quad \frac{n!}{(n-k)! \cdot k!}.$$

For example, the third entry, $k = 3$, in the fifth row, $n = 5$, is $5!/(2! \cdot 3!) = 10$. We note that $0! = 1$. Some people may not like this but it really simplifies things (and there are some excellent arguments in favor of setting $0! = 1$). This formula is so important that it gets its own symbol, called the 'binomial coefficient'

$$\binom{n}{k} = \frac{n!}{(n-k)! \cdot k!}$$

One can think of the symbol on the left-hand side as an *abbreviation* for the right-hand side. If things have their own abbreviation, it is usually a sign that they are important (or, at the very least, that they show up frequently, which is maybe the same thing). If the Pascal Triangle has many magic properties and if the Pascal Triangle is made up of binomial coefficients $\binom{n}{k}$ it stands to reason that $\binom{n}{k}$ also has a lot of interesting properties. The most important one is $\binom{n}{k} = \binom{n-1}{k-1} + \binom{n-1}{k}$ which corresponds to 'the k-th entry in the n-th row is the sum of the two entries above' which is exactly the way we defined Pascal's Triangle. It is not too difficult to understand why this equation is true: using the definitions

$$k! = k \cdot (k-1)! \qquad \text{as well as} \qquad (n-k)! = (n-k) \cdot (n-k-1)!$$

we can plug these in and get

$$\begin{aligned}
\binom{n-1}{k-1} + \binom{n-1}{k} &= \frac{(n-1)!}{(n-k)! \cdot (k-1)!} + \frac{(n-1)!}{(n-1-k)! \cdot k!} \\
&= \frac{(n-1)! \cdot (k + (n-k))}{(n-k)! \cdot k!} = \frac{n!}{(n-k)! \cdot k!} = \binom{n}{k}.
\end{aligned}$$

Binomial coefficients do not only show up as numbers in the Pascal triangle, they also have a completely separate definition:

$\binom{n}{k}$ is the number of ways of choosing k elements from a collection of n elements.

For example, if we have the first 4 letters a, b, c, d and we look at all possible ways of choosing 2 (different) letters, we find that we can choose ab, ac, ad, bc, bd, cd which are 6 possibilities and indeed $\binom{4}{2} = \frac{4!}{2! \cdot 2!} = \frac{24}{2 \cdot 2} = 6$. This also give us a good reason why it might make sense to have $0! = 1$. How many ways are there to choose n things from a collection of n things? Well, there is really only one way to do it: pick all of them. Thus, if we want the equation to still be true, we have to have $1 = \binom{n}{n} = \frac{n!}{(n-n)! \cdot n!} = \frac{n!}{0! \cdot n!} = \frac{1}{0!}$ and this shows that $0! = 1$. This *combinatorial* way of thinking about the binomial coefficient, thinking about it as counting k-element subsets of a set with n elements, is frequently quite useful. We can use it to prove the symmetry of the Pascal Triangle. Formally, we want to show that $\binom{n}{k} = \binom{n}{n-k}$. The number of ways of picking k things from a collection of n things is exactly the same as the number of ways of picking $n - k$ things from a collection of n things: for each way of picking k out of n things, we may simply look at the things that remain, that were *not* picked, and their number is exactly $n - k$. There is a 1-to-1 correspondence between the set of k-element subsets and the set of $(n - k)$-element subsets of a set of n elements, and therefore they have to have the same number. Having the binomial coefficients at our disposal also allows us to write down a general form of the binomial theorem

$$(1+x)^n = \binom{n}{0} + \binom{n}{1}x + \binom{n}{2}x^2 + \cdots + \binom{n}{n-1}x^{n-1} + \binom{n}{n}x^n.$$

This is not a coincidence! Suppose we try to multiply $(1 + x)^7 = (1 + x)(1 + x)(1+x)(1+x)(1+x)(1+x)(1+x)$. One way of multiplying it out is to do it as one learns in High School, one parenthesis at a time. Of course the arising expressions get complicated very quickly. Another way of going about it is to ask how we might end up multiplying things to get a term like x^3: it would mean that, when multiplying out, we have to pick x three times and 1 four times. How many ways are there to choose 3 parentheses from a list of 7? The answer is exactly $\binom{7}{3}$ and that is the corresponding number in the binomial theorem.

2.5 A detour: Notation

This is an excellent place to quickly talk about mathematical notation: why do we use $+$ to denote addition and $-$ to denote subtraction? Or, given the previous section, why do we use $\binom{n}{k}$ to denote the binomial coefficient? Arguably, there are many other ways we could write it: one could write

$$\Theta^n_k \qquad \text{or maybe} \qquad \bigoplus n \otimes k \bigoplus \qquad \text{or perhaps} \qquad \boxed{n\,|\,k}.$$

In the case of binomial coefficients, we use $\binom{n}{k}$ because Andreas von Ettingshausen (1796–1878), when writing his 1826 book *The Combinatorial Analysis as a preperatory course for the study of higher mathematics*, said that

> Since in subsequent matters we will very frequently have the chance to make use of the numerical expression of this set, we will use the symbol $\binom{n}{k}$ to denote it. (Andreas von Ettingshausen, *The Combinatorial Analysis as a preperatory course for the study of higher mathematics*, 1826, [79])

Ettingshausen explicitly writes that he is *choosing* the symbol $\binom{n}{k}$. It is a choice, it is arbitrary and it happened to catch on! Notation is completely arbitrary: one could always write things in a different way. This is even true for (non-mathematical) language. Most of us speak at least one language, we usually do not remember how we learned that one and it is not so easy to explain how we speak it: it just kind of works? The problem of how language works is quite challenging and, as is frequently the case with challenging problems, the early treatment is quite often particularly charming (see Fig. 2.7).

Fig. 2.7: Ferdinand de Saussure's *Course in general linguistics* [252].

The Swiss linguist Ferdinand de Saussure (1857–1913) made an explicit distinction between the *signifier* (the word, the symbol) and the *signified* (that which the word refers to). The word 'tree' (signifier) refers to the plant 🌳 (signified). A natural question is how these two concepts, the signifier and the signified, are connected and de Saussure leaves absolutely no room for doubt

> The bond between the signifier and the signified is arbitrary. Since I mean by sign the whole that results from the associating of the signifier with the signified, I can simply say: *the linguistic sign is arbitrary.* (Ferdinand de Saussure, *Course in general linguistics* [252])

He goes on to explain that *The idea of "sister" is not linked by any inner relationship to the succession of sounds s-ö-r which serves as its signifier in French; that it could be represented equally by just any other sequence is proved by differences among languages* [252]. Words, symbols and signs are not fundamentally linked to what they represent, it is a matter of convention. De Saussure lists one amusing exception to the rule: *onomatopoeia*, words that phonetically resemble what they represent ('tic-toc' for the sound of the clock, 'glug-glug' for the sound of drinking, 'meow' for the cat). The physicist Richard Feynman (1918–1988) gives a nice example.

> When I was younger I was anti-culture, but my father had some good books around. One was a book with the old Greek play *The Frogs* in it, and I glanced at it one time and I saw in there that a frog talks. It was written as "brek, kek, kek." I thought, "No frog ever made a sound like that, that's a crazy way to describe it!" so I tried it, and after practicing it awhile, I realized that it's very accurately what a frog says. (Richard Feynman [83])

Returning to mathematical notation, the signs $+, -, \binom{n}{k}$ are all *signifiers*: they represent an underlying idea like addition, subtraction or the binomial coefficient (the *signified*). Following de Saussure, the signifier is arbitrary and could be equally replaced by something else, there is no intrinsic fundamental connection. However, de Saussure continues, changing things is not so easy: *The signifier [...] is fixed, not free, with respect to the linguistic community that uses it. The masses have no voice in the matter. [...] No individual [...] could modify in any way at all the choice that has been made* [252]. Feynman made the same discovery:

I didn't like the symbols [...] so I made a different sign, something like an &. [...] I thought my symbols were just as good, if not better, than the regular symbols – it doesn't make a difference what symbols you use – but I discovered later that it *does* make a difference. Once when I was explaining something to another kid in high school, without thinking I started to make these symbols, and he said, "What the hell are those?" I realized then that if I'm going to talk to anybody else, I'll have to use the standard symbols. (Richard Feynman [83])

2.6 π and Singmaster

We finish our discussion of Pascal's triangle with one last property and one vexing question. If there is an odd number of numbers showing up in a row of Pascal's triangle, then we can speak of the 'middle' element. These 'middle' numbers are given by the formula

$$\frac{(2n)!}{(n!)^2} = \binom{2n}{n}$$

and the first few are $1, 2, 6, 20, 70, 252, 924, \ldots$ We see that these numbers grow pretty quickly and one could now ask: how quickly? A careful look suggests very fast (exponential) growth, and this is correct, these numbers grow just a little bit slower than 4^n (but much faster than 3.99^n). The more precise answer, however, is shocking: when n gets larger and larger, then

$$\frac{\sqrt{n}}{4^n} \cdot \left(\frac{(2n)!}{(n!)^2}\right) \qquad \text{gets closer and closer to} \qquad \frac{1}{\sqrt{\pi}}.$$

To illustrate this, if we set $n = 1000$, then the expression is $0.564119\ldots$ while $1/\sqrt{\pi} = 0.56419\ldots$. This is a pretty good approximation already – and it gets even better when n becomes larger. π is a purely geometric quantity, it is the area of the unit disk! How is it possible that Pascal's triangle, a rule about adding integers, knows about π? Even though we know many things about Pascal's triangle, there are also some things we do not know. One very funny question is *Singmaster's problem*. It is named after David Singmaster (1938–2023), who demonstrates that, even today, mathematicians are not necessarily doomed to live a boring life! He started by getting kicked out of college, *I went to college at Caltech and at the end of my third year I got thrown out for lack of academic ability* [265]. He did not get discouraged: after getting a PhD at Berkeley, he moved to London in 1970 where he would stay for the rest of his life – but not exclusively. In 1971 he joined an expedition to Sicily, Italy, organized by one of the pioneers in underwater archeology, Honor Frost. There, by *lack of discipline*, he made a wonderful discovery. Honor Frost recalls

Lack of discipline in this "swim-line technique" lead to the discovery of the "Punic Ship". David Singmaster (expedition photographer in 1971) signalled a find. Considering him to be off-course and too near to the shore to have come upon anything significant I reluctantly answered his call. The mysteries of the sea are, however, inexhaustable; between two piles of ballast-stones, a large timber (such as I had never seen before) emerged from the sand like the head of a primaeval animal crowned with weed; the presence of a buried wreck was evident. (Honor Frost, *The Punic Ship: Final Excavation Report*, [94])

This was an important discovery: *No ancient warship had ever been found and it seemed improbable that this would ever occur in the sea* [94]. The Marsala Punic Ship is believed to have been involved in the *Battle of the Aegates* near the Aegates Islands in 241 BC, the final and deciding battle of the First Punic War between the Roman Republic and Carthage. So it seems that 1971 was a pretty good year for David Singmaster, but that is not where the story ends: besides discovering an ancient warship he also found a really great question: how often can a number appear in the Pascal triangle? The number 1 appears infinitely many times (twice in each row). The number 2 appears once, the numbers 3,4,5 appear twice, the number 6 appears 3 times. The current record is the number 3003 which appears 8 times

$$3003 = \binom{3003}{1} = \binom{78}{2} = \binom{15}{5} = \binom{14}{6}$$
$$= \binom{14}{8} = \binom{15}{10} = \binom{78}{76} = \binom{3003}{3002}.$$

It seems quite plausible, for a variety of reasons, that this may indeed be the record. Maybe there is another number that we have not found yet that appears 10 times or maybe even 12 times but at this point it seems difficult to believe that there is a number larger than 1 that appears more than 1000 times – on the other hand, we do not know for sure and there are a lot of numbers!

2.7 Exercises

1. Write down the first 10 rows of Pascal's triangle by hand.
2. Go through each row and sum all the entries in that row. Do you notice any pattern? Maybe the Binomial Theorem can help in finding an explanation.
3. Come up with 5 different examples of onomatopoeia!
4. Prove that $2^{n-1} \le n! \le n^n$. Can you prove that for all $c \ge 1$ and all n sufficiently large (depending on c) one has $n! \ge c^n$?
5. Prove the two equations

$$\binom{n}{k} = \frac{n}{k}\binom{n-1}{k-1} \quad \text{and} \quad \binom{n-1}{k} = \frac{n-k}{n}\binom{n}{k}.$$

6. One beautiful fact (taken from Nelson [210]) involving binomial coefficients is

$$(1^1 \cdot 1!) \cdot (2^2 \cdot 2!) \cdot (3^3 \cdot 3!) \cdots (n^n \cdot n!) = (n!)^{n+1}.$$

   Try to prove it! If it turns out to be too difficult, wait until after §6.
7. Prove that $\frac{1}{n+1}\binom{2n}{n} = \binom{2n}{n} - \binom{2n}{n+1}$. These numbers are known as the *Catalan numbers*, they show up everywhere!
8. Wilson's Theorem says that $(n-1)! + 1$ is divisible by n *if and only if* n is a prime number. This statement has two parts: (1) if n is a prime number, then

$(n-1)!+1$ is divisible by n and (2) if n is not a prime number, then $(n-1)!+1$ is not divisible by n. See whether you can find an explanation for (2).

9. One of the fun things in mathematics is that you can always change the rules. Try to come up with your own triangle summation rule. For example, instead of summing the two entries above, one could sum four entries above (replacing things by 0 if they are not defined). Any fun new discoveries?

10. Discuss whether Stifel's numerology $a = 1, b = 2, c = 3$ is a formal system. What about his second system $a = 1, b = 3, \ldots$? Is it part of mathematics?

11. Michael Stifel was not the only mathematician to predict the end of the world.

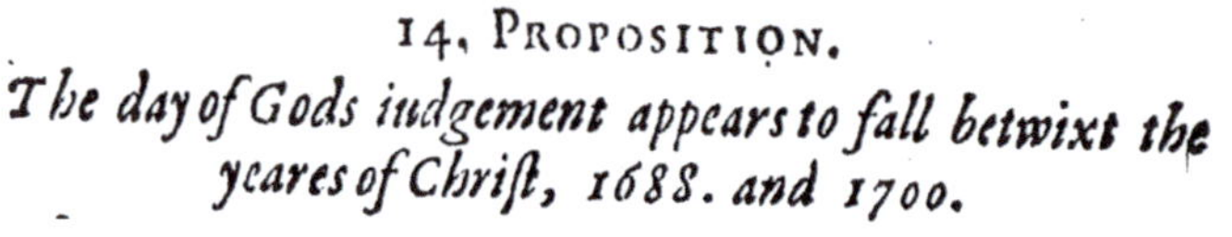

14. PROPOSITION.

The day of Gods iudgement appears to fall betwixt the yeares of Chriſt, 1688. and 1700.

Fig. 2.8: Napier's 1593 *A Plaine Discovery* [206].

One of them was the Scottish mathematician John Napier (1550–1617), who invented *logarithms*, but was particularly proud of his work [206] which has the long title *A plaine discouery of the whole Reuelation of Saint Iohn set downe in two treatises: the one searching and prouing the true interpretation thereof: the other applying the same paraphrastically and historically to the text. Set foorth by Iohn Napeir L. of Marchistoun younger. Whereunto are annexed certaine oracles of Sibylla, agreeing with the Reuelation and other places of Scripture.* For history's sake, it should be mentioned that Napier was not the first to invent logarithms. Jost Bürgi (1552–1632) was a Swiss clockmaker who may have invented logarithms, in a different and less structured way than Napier, for his own personal use. Johannes Kepler (1571–1630) gives a colorful description of this.

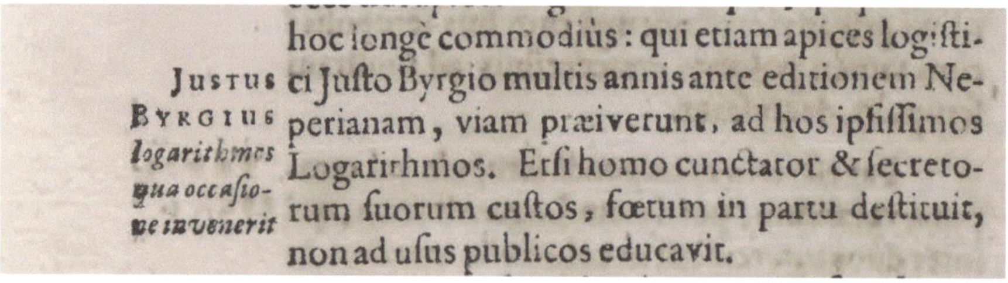

> hoc longe commodiùs : qui etiam apices logiſti-
> JUSTUS ci Juſto Byrgio multis annis ante editionem Ne-
> BYRGIUS perianam, viam præiverunt, ad hos ipſiſſimos
> *logarithmcs* Logarirhmos. Etſi homo cunctator & ſecreto-
> *qua occaſio-* rum ſuorum cuſtos, fœtum in partu deſtituit,
> *ne inuenerit* non ad uſus publicos educavit.

Fig. 2.9: Johannes Kepler in his 1627 *Rudolphine Tables* [153] discussing the work of Jost Bürgi: *Justus Byrgios, many years before the Neperian Edition, led the way to these logarithms. However, he was a hesitant man and the keeper of his secrets, he abandoned the baby at birth and did not raise it for public use.*

Have a look at the predictions for the end of the world, there are many!

Chapter 3
Some Simple Number Mysteries

The Pascal triangle is a good example of a formal structure where we can understand some things but also find some behavior that we do not understand (at least not yet). The same is true for the good old integers, a formal system that we know very well. One might be inclined to believe that we know almost everything about the integers, but there are some shockingly simple things that remain not understood. We will discuss three examples (and many more will appear later).

3.1 The 196 mystery

Here is a fun little game: take any integer, for example 23. We then add to the number the reverse of its digits ($23 \rightarrow 32$)

$$23 + 32 = 55.$$

We see that 55 is the same number read forwards and backwards (a palindrome). It is easy to see that this process does not always lead to a palindrome. For example, if we take 19, then its reverse is $19 \rightarrow 91$ and $19 + 91 = 110$ is not a palindrome. However, repeating the process one more time, $110 + 011 = 121$, we end up with a palindrome. Sometimes even more steps are required, a good example is 79

$$79 + 97 = 176$$
$$176 + 671 = 847$$
$$847 + 748 = 1595$$
$$1595 + 5951 = 7546$$
$$7546 + 6457 = 14003$$

and then, finally, $14003 + 30041 = 44044$. So far, this may seem like a good way to pass the afternoon and have some fun with palindromes along the way, that is, until one tries the number 196. The number 196 leads to the sequence

$$196 \rightarrow 887 \rightarrow 1675 \rightarrow 7436 \rightarrow 13783 \rightarrow 52514 \rightarrow 94039 \rightarrow 187088 \rightarrow \ldots$$

People have computed millions and millions of steps and never found a palindrome. The question is now: will this procedure, started with 196, ever produce a palindrome? There are good reasons to believe that this will never happen, but we do not know!

© The Author(s), under exclusive license to Springer Nature Switzerland AG 2025
S. Steinerberger, *The Unreasonable Elegance of Mathematics*,
Springer Undergraduate Mathematics Series, https://doi.org/10.1007/978-3-032-03815-9_3

3.2 Interlude: why do we care?

The 196 mystery is maybe a good point to stop and ask: why do we care? It may be fun to produce palindromes and it may also be fun to wonder what makes 196 special. However, few would be willing to spend the next year of their life trying to uncover the 196 mystery: so why do mathematicians care? The answer is simple: the vast majority of mathematicians do not. Just as avid readers do not love every book, mathematicians are not interested in every single math problem – and very much like readers who usually have a preference for a certain book genre (true crime, science fiction, ...), most mathematicians tend to be interested in a particular flavor of problem (which may be geometric, combinatorial, algebraic, analytical, ...).

There is a second conundrum that is hidden here: the conundrum of *basic science*. If someone wanted to spend the rest of their life trying to unlock the 196 mystery, well, surely we would wish them the best of luck. Imagine now the same person asked to be *paid* to study the 196 mystery (say, by occupying a position at a University or as part of a government research lab). There might be an instinctual reaction of 'well, isn't there a better use of resources'? Some people embrace the uselessness.

> I have never done anything useful. No discovery of mine has made, or is likely to make, directly or indirectly, for good or ill, the least difference to the amenity of the world. I have helped to train other mathematicians, but mathematicians of the same kind as myself, and their work has been, so far at any rate as I have helped them to it, as useless as my own. Judged by all practical standards, the value of my mathematical life is nil; and outside mathematics it is trivial anyhow. (G. H. Hardy [125])

This leads to an interesting aspect of mathematical history: it has happened many times that 'obviously' useless questions end up being of significant usefulness in real life. Greek mathematicians studied ellipses because they thought they were amusing (much like the 196 mystery) until Kepler discovered (millennia later) that planets might be moving on ellipses. In the case of G. H. Hardy and the supposed uselessness of his work and that of his students, the list of his students includes

1. Sidney Chapman, who started his research career with Hardy, and who first described, in *On Ozone and Atomic Oxygen in the Upper Atmosphere* [52], how the ozone layer is created!

2. I. J. Good, who wrote an entire essay on *Speculations Concerning the First Ultraintelligent Machine*, where he raises an important point

   > Let an ultraintelligent machine be defined as a machine that can far surpass all the intellectual activities of any man however clever. Since the design of machines is one of these intellectual activities, an ultraintelligent machine could design even better machines; there would then unquestionably be an 'intelligence explosion,' and the intelligence of man would be left far behind... [...] It is sometimes worthwhile to take science fiction seriously. (I. J. Good [114])

3. Edward Linfoot, who exhibited a *microscope of his own construction, became Assistant Professor of the University Observatory* and *was engaged in the early stages of the Edsac I, one of the first generation of fast computers* [173].

4. Ethel Newbold, who was tutored by Hardy while she was a student. Her 1927 paper *Practical Applications of the Statistics of Repeated Events, Particularly to Industrial Accidents* specifically says that

> The main questions that the originators of these investigations had in mind were (1) Does any definite tendency exist [...] for certain people to sustain more accidents than others? [...] (3) If such a tendency exists, to what extent can it be measured, and does it show any association with other qualities? (4) Is it possible to devise any sort of test by which people liable to accidents, if such exist, could be roughly sorted out, so that in the choice of persons for occupations to which special risk is attached such individuals could be avoided? (Ethel Newbold [214])

This work was awarded the Guy Medal of the Royal Statistical Society; since 2014, the Bernoulli Society has awarded the *Ethel Newbold Prize*.

In the case of Hardy himself, he has done seminal work on prime numbers $(2, 3, 5, 7, \dots)$, which are now the basis of modern cryptography: secure communication is secure as long as it is very difficult to decompose a very large integer into a product of prime numbers (which we believe to indeed be hard). So one might well wonder whether a resolution of the 196 mystery will, in a couple of hundred years, form the basis of time travel technology. The author is not necessarily convinced that this will be the case, but the author has been wrong before.

3.3 Ulam numbers

> Ulam did not make it through the war years unscathed. In the winter of 1945 he was rushed to a hospital in Los Angeles, suffering a violent headache, numbness in his chest, and slurred speech. [...] The doctors found that Ulam's brain was severely inflamed, "bright pink instead of the usual gray." They gave him antibiotics and drilled a hole in his skull to relieve the pressure. When the swelling subsided and Ulam awoke after a few days of postoperative coma, his surgeon tested his mental faculties by asking him the sum of 8 and 13. "The fact that he asked such a question embarrassed me so much that I just shook my head," recalled Ulam. "Then he asked me what the square root of twenty was, and I replied: about 4.4. He kept silent, then I asked, 'Isn't it?' I remember Dr. Rainey laughing, visibly relieved, and saying, 'I don't know.' " (Paul Hoffman, *The Man Who Loved Only Numbers* [138])

In 1964, Stanislaw Ulam, for reasons that are not entirely clear, discussed a particular sequence of integers [287]. We do not know *why* he was interested in the sequence in the first place, the original description is as follows.

> Another one-dimensional sequence was studied recently. This one was defined purely additively: one starts, say, with integers 1,2 and considers in turn all integers which are obtainable as the sum of two different ones previously defined but only if they are so expressible in a unique way. The sequence would start as follows:

$$1, 2, 3, 4, 6, 8, 11, 13, \dots;$$

> even sequences as simple as that present problems! (Ulam [287])

Let us first discuss very carefully what he means. The rules are as follows: we start with the numbers 1 and 2. Then we add the smallest number that can be uniquely written as the sum of two distinct earlier terms. That would be 3, since $3 = 1 + 2$. We have $1, 2, 3$. The next number is 4 because $4 = 1 + 3$ (someone might say that it's also $4 = 2 + 2$, but we only care about sums of distinct earlier terms, one cannot use an existing number twice). So now it's 1,2,3,4. The next number is *not* 5 since $5 = 1 + 4 = 2 + 3$ cannot be written as a unique sum, there are different ways of writing it. But 6 is fine since $6 = 2 + 4$ and that's the only way. Continuing like this, we end up with

$$1, 2, 3, 4, 6, 8, 11, 13, 16, 18, 26, 28, 36, 38, 47, 48, 53, 57, \ldots$$

Should we care? It looks like an arbitrary rule and a pretty arbitrary sequence. Very few people actually cared, it does not look very interesting. Many years later it was discovered that the sequence appears to have a strange and unexpected self-organizing property and that makes it very interesting. First, we know that there are infinitely many numbers in the Ulam sequence. The reason is actually quite simple: suppose we compute the first few terms and end up with $1, 2, 3, 4, 6, 8, 11, 13, 16, 18, 26$. How do we know that the sequence will continue? Well, if the sequence were to stop here, then the number $18 + 26 = 44$ is guaranteed to be a number that can be written as the sum of two terms in a unique way: any other sum of two other terms would be smaller because we picked the two largest numbers. This does not mean that the next element will be 44 (and indeed it is not, the next element is 28) but it means that if none of the number $27, 28, \ldots, 43$ are part of the sequence, then 44 would be a way to continue. This argument actually shows a little bit more, it shows that the sequence does not grow faster than exponentially. This leads to our first 'Theorem'. The term 'Theorem' is a fancy way of saying 'True Statement'. Like many things in mathematics, it comes from Greek and means something close to 'I observe a point of view' (*theory* and *theater* have the same origin).

Theorem. *The n-th Ulam number a_n satisfies*

$$a_n \leq 2^{n-1}.$$

Looking back at the first few terms, it seems that this upper bound is *much* larger than the actual behavior. Doing some more experiments, one observes that the n-th Ulam number seems to be approximately $a_n \sim 13.5n$ for reasons that are not understood. To make things even stranger, it seems [267] that if a_n is a number appearing in the Ulam sequence, then we have

$$\cos\left(2.57145...a_n\right) < 0 \qquad \text{with only four exceptions} \quad 2, 3, 47, 69.$$

What is this strange number 2.57145... and where does it come from? This phenomenon has now been tested on a computer for many millions of terms but remains very poorly understood. Did Ulam know about this?

3.4 The Collatz Problem

"Mathematics may not be ready for such problems." (attributed to Paul Erdős)

Be warned: many people who learn about this problem end up becoming obsessed. The problem was first introduced by the mathematician Lothar Collatz. A short biography of Collatz writes with a regretful undertone *In many ways it might seem a pity that a mathematician who has produced so much important and fundamental work should be most remembered for a novelty, yet this problem has intrigued mathematicians ever since he proposed it in 1937* [185]. The problem is tremendously simple: we introduce two rules

1. If n is even, replace it by $n/2$.
2. If n is odd, replace it by $3n + 1$.

So, for example, if we start with 18, we get

$$18, 9, 28, 14, 7, 22, 11, 34, 17, 52, 26, 13, 40, 20, 10, 5, 16, 8, 4, 2, 1$$

and after reaching 1, the rules always produce $1, 4, 2, 1, 4, 2, 1, \ldots$.

Question. Does this process always reach 1 independently of which number we start from?

We note that we can simplify the process a little: every time n is odd, we note that $3n$ is also odd and $3n + 1$ is even. But if it's even, then we are going to divide it by 2, so we can make the process go a little bit faster.

1. If n is even, replace it by $n/2$.
2. If n is odd, replace it by $(3n + 1)/2$.

The associated question is simple: no matter with which integer we start, if we keep applying the rules, will we eventually end up at 1?

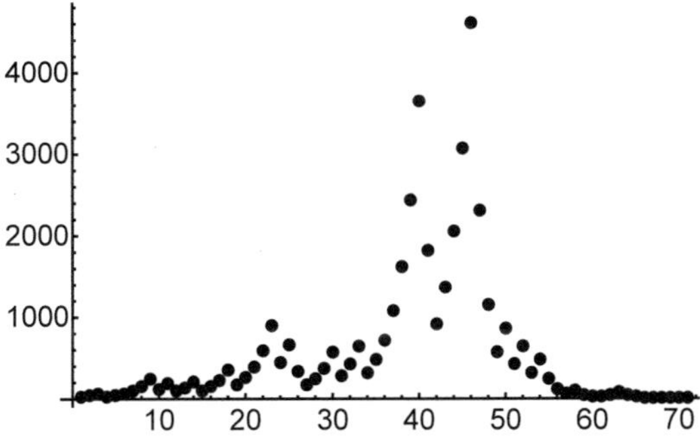

Fig. 3.1: The Collatz procedure applied to 27.

What is strange about the process is that it is very hard to predict. So, for example, the sequences $26, 13, 20, 10, 5, 8, 4, 2, 1$ and $28, 14, 7, 11, 17, 26, 13, 20, 10, 5, 8, 4, 2, 1$ are fairly short, but if we start with 27, we get a list of 71 numbers, the largest of which is 4616. Truly unpredictable! What makes this problem so attractive? One aspect is surely that it is a completely elementary question about addition and multiplication; there is a certain feeling of 'we should be able to figure this out'. However, there are other factors: one reason why people are frustrated is because there exists a very simple reason that explains very well why every number should eventually go down to 1.

> Half the numbers are even and half are odd. If we divide by 2, we multiply by 0.5, otherwise we multiply by 1.5. But, on average, we multiply by $\sqrt{0.5 \cdot 1.5} \sim 0.866 < 1$ and so, on average, things should go down to 1 pretty quickly.

There is no trick here, this little argument appears to correctly predict the behavior of the sequence fairly accurately. This is in contrast to the example of 27 above, where we were unable to predict even the most basic things. However, this is partially because 27 is still a pretty small number. If we start with a ridiculously large number, say $10^{100} + 1$, more regularities emerge: $10^{100} + 1$ has 100 digits and needs 1462 steps to go down to 1. If we simply plot the *number of digits* that the numbers that we see along the evolution, we get the following picture. For comparison we also

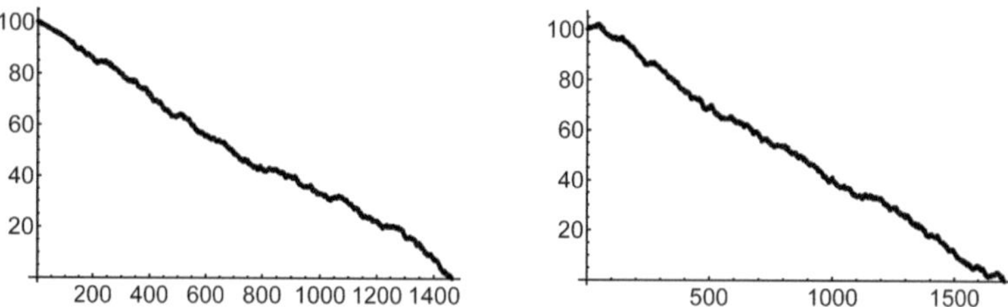

Fig. 3.2: The Collatz procedure applied to $10^{100} + 1$ (left) and $10^{100} + 27$ (right).

plot the process for $x = 10^{100} + 27$, which is a nearby starting value with a very different evolution: it needs 200 extra steps to reach 1, the maximum value attained is more than 100 times larger than the one attained when starting with $x = 10^{100} + 1$. However, the number of digits appears to be fairly similar, there is linear downward trend. We could now wonder what that decay rate $0 < q < 1$ might be. Taking the figure on the left, we go from $10^{100} + 1$ down to 1 in ~ 1500 steps which suggests the equation $(10^{100} + 1) \cdot q^{1500} = 1$ resulting in the solution $q \sim 10^{-1/15} \sim 0.857$ which is not too far from $\sqrt{0.5 \cdot 1.5} \sim 0.866$ predicted above. This is where things get annoying: assuming things to be random leads to a very accurate description of what seems to happen. However, nothing here is random, everything is precisely pre-determined. Starting with 27 we will always need 71 steps and the largest number we encounter along the way will always be 4616.

3.5 Mathematical diseases

First you will come to the Sirens who enchant all who come near them. If any one unwarily draws in too close and hears the singing of the Sirens, his wife and children will never welcome him home again, for they sit in a green field and warble him to death with the sweetness of their song. There is a great heap of dead men's bones lying all around, with the flesh still rotting off them. (Homer, *Odyssey* [142])

Such problems have been called **mathematical diseases**, a terminology that goes back to Frank Harary (1921–2005). Harary was a well-traveled mathematician, so well-traveled that he *gave lectures in cities with names beginning with every letter of the alphabet. For example, he lectured in Xenia, Ohio, in the United States and in Xanten, Germany* [53]. According to Harary, just like the flu, a mathematical disease is easily caught. To qualify as a mathematical disease, a problem needs to be

1. Easy to state, preferably using only elementary concepts.
2. It needs to appear accessible, almost as if one extremely simple idea would be enough to solve it.
3. Finally, there have to be many incorrect solutions. The problem has already attracted a lot of attention with no noticeable progress.

People who are interested in learning more about the Collatz problem should reconsider their life choices: however, if they have already caught the disease, they should start with the book of Lagarias [161].

Come here, they sang, *renowned Ulysses, honor to the Achaean name, and listen to our two voices. No one ever sailed past us without staying to hear the enchanting sweetness of our song – and he who listens will go on his way not only charmed, but wiser, for we know all...* They sang these words most musically, and as I longed to hear them further I made by frowning to my men that they should set me free; but they quickened their stroke, and Eurylochus and Perimedes bound me with still stronger bonds till we had got out of hearing of the Sirens' voices. Then my men took the wax from their ears and unbound me. (Homer, [142])

3.6 Exercises

1. Play the 196 game with starting value 89. It takes a while but, I promise, it does not take more than 30 steps.
2. Learn about the life of Stanislaw Ulam. He had many great ideas (another of which will be discussed in §12) and a very interesting life!
3. Compute the first 20 elements of the Ulam sequence by hand. It is quite easy in the beginning and then gets more complicated: one has to check many combinations.

4. An Ulam number can be written as the sum of two distinct earlier Ulam numbers in a *unique* way. That is part of the definition. A non-Ulam number can often be written as the sum of two Ulam numbers in *many* different ways. The Ulam numbers start $1, 2, 3, 4, 6, 8, 11, 13, 16, 18, 26, 28, 36, 38, 47, 48, 53, 57, \ldots$ We can write 53, an Ulam number, uniquely as the sum of two earlier terms, $53 = 6 + 47$. 54 is not an Ulam number, so it can be written in more than one way as the sum of two earlier terms: in fact, there are many more such representations

$$54 = 1 + 53 = 6 + 48 = 16 + 38 = 18 + 36 = 26 + 28.$$

Explore why that might be!

5. Prove that there are infinitely many numbers n that, under the Collatz procedure, will eventually end up at 1. (The Collatz conjecture says that all numbers n will eventually lead to 1, that is a much stronger statement).

6. If we keep applying the Collatz rules even after we reach 1, we enter into the cycle $1 \rightarrow 4 \rightarrow 2 \rightarrow 1 \rightarrow 4 \rightarrow \ldots$ and stay in this cycle forever. Prove that no other cycle of length 3 exists. Can you prove that no cycle of length 4 exists?

7. How many steps does $10^{10000} + 1$ need to reach 1 with Collatz rules? (A computer may be useful here). Can you use the number of steps to predict the average geometric decay rate? Is it close to $\sqrt{3/4} \sim 0.866$?

8. A stronger form of the Collatz conjecture is that not only does starting in the integer n eventually lead to 1, it is also believed that this happens within less than $100 \log(n)$ steps. See whether you can find a reason why one might believe this.

A warning (xkcd.com/710)

Chapter 4
Cellular Automata

4.1 People moving into houses

We defined mathematics as the study of formal structures. This is quite general and includes things that we really do *not* understand. The purpose of this section is to illustrate a simple type of game that quickly produces results that are quite mysterious. Assume that there are infinitely many little squares (that we think of as uninhabited houses) next to each other, maybe like this

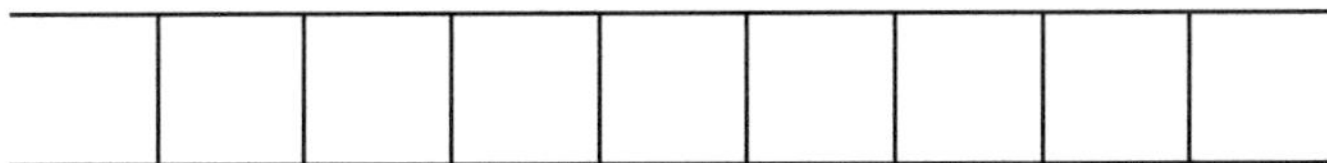

Assume furthermore that all these houses are empty except for a single house somewhere that contains a happy person. It is not important that the person is happy, all people in this example are happy.

The rules of the game are now as follows

1. If a house has exactly one neighboring house that is inhabited, then a person moves in (it's nice to have a neighbor).
2. If a house is inhabited and surrounded by two houses that are occupied, then the person moves out (too much noise).

Let us see how the game evolves: in the beginning, there is a single house that is inhabited. There are two houses next to it and these are attractive real estate: there is a nice neighbor nearby. People are going to move in and it will look like

There are now three occupied houses next to each other and two things happen: the two houses to the left and to the right of the group of three are empty and somebody will move in (it is nice to have a neighbor) but the person in the middle house will move out (too much noise). Therefore the new situation will be as follows

© The Author(s), under exclusive license to Springer Nature Switzerland AG 2025
S. Steinerberger, *The Unreasonable Elegance of Mathematics*,
Springer Undergraduate Mathematics Series, https://doi.org/10.1007/978-3-032-03815-9_4

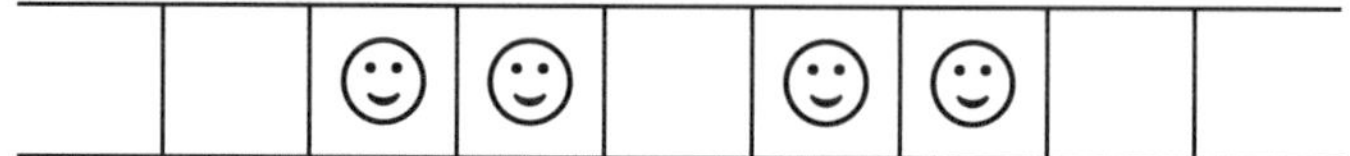

We see again, just as before, there are the two houses bordering on the leftmost house and the rightmost house that have exactly one neighbor, people will move in. The empty house in the middle stays unoccupied because there are two neighbors (too much noise). Nobody is surrounded by two neighbors, so nobody moves out:

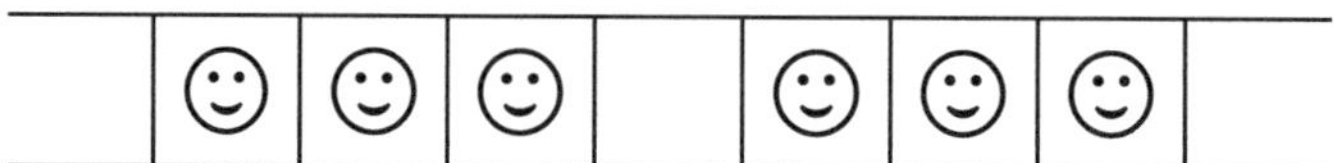

At this point, two more neighbors are going to move in and two are moving out (too much noise) and the new situation will be

It makes sense to change the way we represent the process. Instead of looking at the houses at a fixed point in time, we visualize the process by looking at the entire evolution. Moreover, for clarity, instead of having little smileys, we color the entire square black if the house is occupied and white if it is not.

This picture leads to some more clarity about what the process is doing: there are always two new houses being added (one on the very left and one on the very right) and the houses that end up in the middle end up being empty. Roughly half the houses will be occupied and half the houses will be empty and they will alternate. If you find this unconvincing, play the game and see what happens!

4.2 Cellular Automata

What we just described is a simple example of a cellular automaton (in fact, its technical name is RULE 94). Summarizing the key ingredients of the game, a cellular automaton can be described by three simple rules.

1. We work with a one-dimensional line of squares (the houses) that are either black (occupied) or white (empty).
2. The next state of the house (whether black or white) depends on its current state and it also depends on the state of the two neighbors (whether the neighboring houses are occupied or not).
3. In order to keep track of the evolution of the system and to visualize it, we show the rows on top of each other.

Since each house is either black or white (2 possibilities) and there are 2 neighboring houses each of which are either black or white (2 possibilities), there are a total of

$$2 \times 2 \times 2 = 8 \qquad \text{different cases.}$$

If we specify what happens in all 8 cases, then we have specified the rules of the game completely. We can test this by returning to the House game from the previous section; the rules were given as a story but we could also list all 8 possible situations and describe what happens in each. This is done in Fig. 4.1.

Fig. 4.1: The Rules of the House Game (RULE 94). The first picture means: if a house is occupied and the two neighboring houses are also occupied, then the person moves out. The third picture means: if a house is empty and the two neighbors are occupied, then the house remains empty.

We can now investigate the behavior of other rules. For any possible black-white pattern one can look at a couple of examples to explore the behavior. Many such rules lead to a nice and somewhat predictable behavior. An example is given by RULE 50 shown in Fig. 4.2.

Fig. 4.2: RULE 50 applied to a single black square.

When we start applying the rules to a single black square, the behavior starts becoming clear. We end up with very nearly the same behavior as in the house game in the previous section. Moreover, once we see what is happening, it is not too difficult to go back to the rules and see *how* they end up producing the pattern that they produce. It all seems easy enough.

4.3 Simple rules lead to simple behavior? RULE 30

The previous two examples, RULE 94 and RULE 50, show that simple rules can lead to simple behavior. This is what one expects: simple rules should lead to simple systems. We phrase this as a guideline that is often used by people both in mathematics but also in other sciences and even everyday life.

Potential Guideline I. Simple rules lead to simple behavior.

This need not always be the case. A fascinating counterexample is the great 1902 rat hunt in Hanoi, Vietnam, where

> a one-cent bounty was paid for each rat tail brought to the authorities (it was decided that the handing in of an entire rat corpse would create too much of a burden). [...] Vietnamese residents began to bring in thousands of tails. [...] officials in the field began to notice a disturbing development. There were frequent sightings of rats without tails going about their business in the city streets. After some perplexity, the authorities realized that less-than-honest but quite resourceful characters were catching rats, but merely cutting off the tails and letting the still-living pests go free (perhaps to breed and produce more valuable tails). (Michael G. Vann [290])

This is an example of a simple rule (one rat tail = one cent) that led to remarkably complicated behavior. However, it is not really a counterexample to Potential Guideline I, indeed, one could argue that the rat story is really the byproduct of remarkably complicated structures that we find in society. The rat-tail rule may be simple but humans are not. Potential Guideline I is also some times used in a reversed form.

Potential Guideline II. Complicated behavior requires complicated rules.

This variant, although perhaps rarely phrased as such, is frequently used. A famous example is William Paley's watchmaker analogy that can be found at the very beginning of his 1802 *Natural Theology*.

> suppose I had found a watch upon the ground [...] the watch must have had a maker: that there must have existed, at some time, and at some place or other, an artificer or artificers who formed it for the purpose which we find it actually to answer; who comprehended its construction, and designed its use. (William Paley, *Natural Theology* [219])

The existence of a watch, argues Paley, is indicative of the existence of a watchmaker. The watchmaker has to be a being complex enough to create the watch in the first place; anything as complicated as a watch has to be created by something of equal or greater complexity. RULE 30 is one example that challenges this belief. RULE 30 is very simple: it is so simple that it can be written out in one line (see Fig. 4.3).

Fig. 4.3: The infamous Cellular Automaton RULE 30.

First described by Stephen Wolfram [304] in the 1980s, it appears to be just another cellular automaton much like RULE 94 (Houses) or RULE 50 (Alternating Squares) that we already saw. Its behavior is anything but simple.

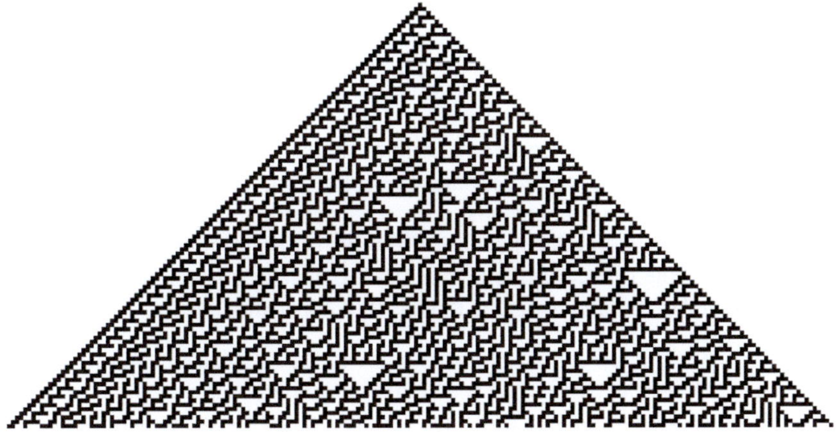

Fig. 4.4: Rule 30: the first 100 steps of the evolution.

Things appear to be fairly regular on the very left side of the picture, we observe some type of repeating 'staircase'. The behavior everywhere else is a little bit more puzzling: there are some fairly large empty triangles and all sorts of other intricate structures. So far, nobody has really understood Fig. 4.4.

Fig. 4.5: *Conus textile*, a sea shell (picture by James St. John).

Similar patterns appear in nature (see Fig. 4.5) for a relatively simple reason: suppose you wanted to remember the first 100 steps in the evolution of RULE 30, basically memorize all of Fig. 4.4. It would take a long time. Alternatively, one could simply memorize the 8 rules that make up RULE 30, that would be much easier, and then recreate Fig. 4.4 by applying the rules. Genetically speaking, a mechanism that iteratively applies a few rules appears to be more likely than a very complicated mechanism that memorizes the entire picture. For more sea shell mathematics, we refer to *The Algorithmic Beauty of Sea Shells* [193].

4.4 The Game of Life

> On a frustrating Friday night circa November 1970, Conway watched a spaceship waltz across
> a screen in the Cambridge Computer Laboratory. [...] Doors closed, lights out, shades drawn,
> Conway was making himself blind, exploring his Game of Life. [...] it was a beautiful invention.
> Initially even Conway would agree with this assessment. Later he had a change of heart.
> (Siobhan Roberts, *Genius at Play: The Curious Mind of John Horton Conway* [243])

Continuing with cellular automata, we now come to its most famous example: the
Game of Life. It is different from the examples we have seen above where we dealt
with cells/squares/houses that are arranged on a line (one-dimensional). In contrast,
the Game of Life is two-dimensional with cells/squares/houses arranged in a grid
structure much like an infinite chess board. For RULE 94, we used a simple story as
a mnemonic for the rules. For the Game of Life, there is a new story: each cell is
either populated or not and the game models the propagation of life.

1. Any live cell with fewer than two live neighbors dies (underpopulation).
2. Any live cell with two or three live neighbors survives.
3. Any live cell with more than three live neighbors dies (overpopulation).
4. Any dead cell with exactly three live neighbors becomes a live cell (reproduction).

Those are some pretty specific rules. Why would anyone possibly come up with
them? Where do these rules come from? John Conway, who invented the Game of
Life in 1970, has a simple answer: hard work.

> I tried out dozens and dozens of different automata, [...] We studied different sets of rules
> [...] over what I think was eighteen months – not all the time, but every now and then during
> coffee breaks. We eventually found this wonderful system. (John Conway [254])

It is worth noting that what we are doing here is Conway's worst nightmare: we are
only mentioning him in the context of the Game of Life. A brilliant and versatile
mathematician with a long history of interesting discoveries, the Game of Life is
what really stuck. As he confided, decades after the discovery, to his biographer

> I told you, every time I turn to a book, a new math book, I look up the sacred name [Conway]
> and it says: 'Conway, Game of Life, pages 34 to 38'. And that's roughly all it says. And how
> can I say it – it's not character assassination, exactly, it's quality assassination. I regard Life
> as trash, frankly. I mean, it was a real part of my life to have discovered it and so on. I don't
> think it should be totally removed. But it seems to be all that I'm known for, among the
> general public. (John Conway [243])

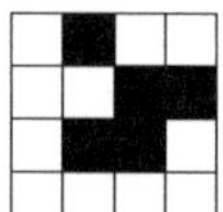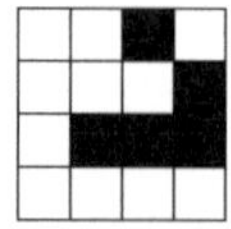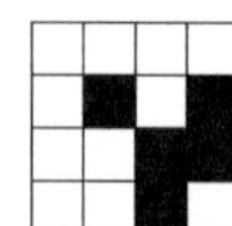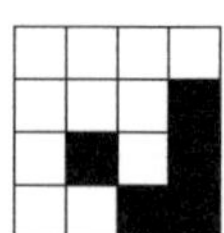

Fig. 4.6: Movement of glider.

One of the things that makes the Game of Life so interesting is that there are incredibly complex patterns. The simplest one, shown in Figure 4.6, is called a *Glider* since it appears to be gliding across the board. It is a simple shape that reproduces itself every 4 steps in a new position: it glides across the board. Configurations that produce an infinite number of gliders have since been discovered (the *Gosper Gliding Gun* being such an example). The rules are simple but they allow for extraordinarily complicated behavior. Many people now actively construct complicated structures in the Game of Life (the Internet being a good source of information and examples). This shows that our simple guideline from above is very wrong.

> **Guideline.** Simple rules can lead to incredibly complicated behavior.

Some people, like the philosopher Daniel Dennett (1942–2024), even go so far as to apply the Game of Life to, appropriately, life itself. Maybe we do not think about it very much but we live in incredibly complicated bodies and have an incredibly complicated interaction with our incredibly complicated environment – quite the miracle! Indeed, it is a miracle for which many explanations have been proposed. Dennett argues that the Game of Life may have something say about this question.

> I use the Game of Life to make vivid for my students the ideas of determinism, higher-order patterns and information. One of its great features is that nothing is hidden; there are no black boxes in Life, so you know from the outset that anything that you can get to happen in the Life world is completely unmysterious and explicable in terms of a very large number of simple steps by small items. No psionic fields, no morphic resonances, no élan vital, no dualism. It's all right there. And the fact that it can still support complex adaptive appropriate structures that do things is also important. (Daniel Dennett [244])

4.5 Exercises

1. Many Cellular Automata, started with a single black square, tend to be pretty boring. Two examples are the rules

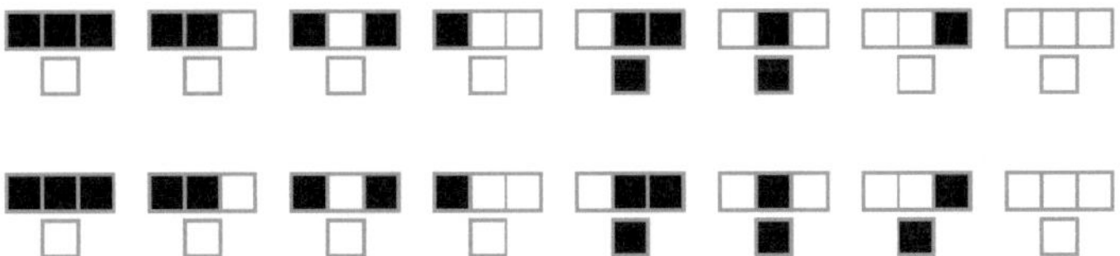

   which are known as RULE 12 and RULE 14. Compute the first few steps in the evolution and see what they do. Once you see the first few steps, go back to the rules and explain what you see.
2. RULE 50 is illustrated in Fig. 4.2. Can you come up with a story behind RULE 50 involving houses and neighbors?
3. Others can be quite fun but not super mysterious. An example is given by RULE 13, which obeys the following law

See what happens to a single black square.

4. We can revisit Pascal's triangle and ask a very simple question: which numbers in the triangle are odd and which numbers are even?

$$
\begin{array}{ccccccccc}
 & & & & 1 & & & & \\
 & & & 1 & & 1 & & & \\
 & & 1 & & 2 & & 1 & & \\
 & 1 & & 3 & & 3 & & 1 & \\
1 & & 4 & & 6 & & 4 & & 1 \\
\end{array}
$$

Fig. 4.7: Odd numbers in the first few rows of Pascal's triangle.

If we identify odd numbers with black squares and even numbers with white squares, we see the rules are actually quite simple: odd+odd = even, even+odd = odd and even + even = even. This can in fact be turned into a cellular automaton that leads to nice behavior. Can you find out what that is?

5. This exercise has three parts that get increasingly more difficult. (1) Go online and watch 3 short videos about Conway's Game of Life. (2) Do not become obsessed and do not let it destroy your life. (3) Do come back to reading the book after watching exactly the 3 videos and nothing else.

6. For one of the cellular automata, RULE 110, Matthew Cook [61] showed that, in principle, every calculation or computer program could be carried out using only RULE 110. Find out what 'Turing complete' means.

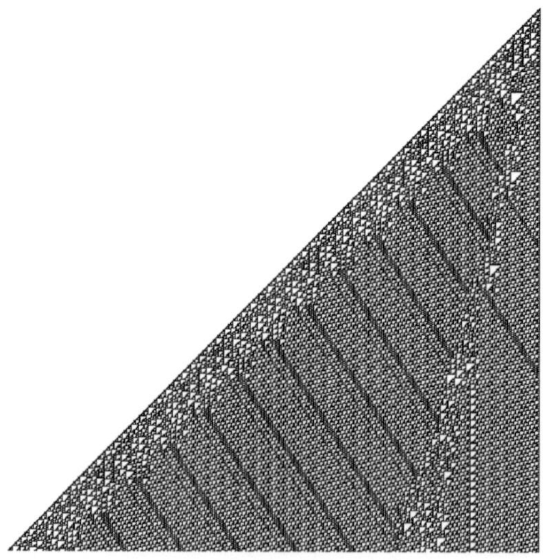

RULE 110

Chapter 5
Axioms: the Rules of the Game

5.1 Rules

In the examples discussed in §4 there were very clear rules that were completely explicit. When playing a game of chess, the rules are also completely clear and can be looked up. Nobody would ever *debate* the rules of chess: the rules of chess are what they are and that is the end of the story. You are free to change some of the rules but that game would no longer be called chess, it would at best be considered a chess variant. The British philosopher Isaiah Berlin gives a colorful description

> *If I ask you 'How do you know that the king in chess moves only one square at a time?' it's no good with someone saying 'Well, you say it moves one square at a time. But, you know, one evening I was looking at the chess board and I actually saw a king move two squares.' This would not be regarded as a refutation of this proposition because what you are really saying is that there is a rule in chess [...]* (Isaiah Berlin [25])

He goes on and explains *and how do you know the rule is true? the rules don't have to be true, they just have to be rules. You either accept them or you don't. This is how mathematics works* [25]. In summary,

1. Mathematics has rules, these are called **Axioms** (like 'Theorem', this is a Greek word that roughly translates to 'that which is worthy of thought').
2. These rules are not debated; if you want to work under the assumption that $1 + 1 = 3$, then you are free to do so but you would not be doing mathematics the way we currently understand it.
3. Until the early 20th century, people were playing fast and loose with the rules/axioms and this had led to various forms of drama and problems. It is worth emphasizing that these problems really only emerged around 1900 and people did *a lot* of mathematics before that (and mostly correctly). Archimedes did a lot of brilliant mathematics without being concerned about the rules/axioms.

Axioms are basic ideas that we use as starting points for reasoning. They are like the building blocks of a system. We accept them without needing proof because they are considered obvious. They are then used to build up the rest. The most famous (and first) example of an axiomatic system goes back to Euclid who described the 'rules' of two-dimensional geometry. The five rules are as follows.

Euclidean axioms

1. A straight line segment can be drawn between any two points.
2. A straight line can be extended indefinitely in both directions.
3. A circle can be drawn with any center and any radius.

© The Author(s), under exclusive license to Springer Nature Switzerland AG 2025
S. Steinerberger, *The Unreasonable Elegance of Mathematics*,
Springer Undergraduate Mathematics Series, https://doi.org/10.1007/978-3-032-03815-9_5

4. All right angles are equal to each other.
5. If a straight line crossing two straight lines makes the interior angles on one side less than two right angles, then the two straight lines, if extended indefinitely, will meet on that side where the angles are less than two right angles.

We will not spend a lot of time with Euclid. However, it is worth mentioning a fascinating historical debate: the fifth axiom (going by the name 'Parallel postulate') is the most complicated one. For a long time people tried to show that it is implied by the other four rules. Maybe there is a way of combining the first four rules in such a way that the fifth rule follows as a byproduct: in that case, we would not need to assume it as a fifth rule, it would be a consequence of the other rules and it could be omitted when listing the rules. We could list the first four rules and then argue that, by suitable combination, they imply the fifth rule as a consequence.

Fig. 5.1: The parallel axiom: for every (infinite) line and every point not on the line there exists a second line going through the point that is parallel to the first.

This turned out to be wrong: there are systems in which the first four rules hold but the fifth does not (non-Euclidean geometries). The history behind this development is a good example of how our understanding of things can progress in the most unusual of ways. Non-Euclidean geometry was developed in 1829–1830 by the Russian mathematician Nikolai Ivanovich Lobachevsky (1792–1856) and, independently, in 1832 by the Hungarian mathematician János Bolyai (1802–1860). It later became clear that Carl Friedrich Gauss (1777–1855) had been studying the problem much earlier. He never published any results because he expected not to be understood:

> I will, for a very long time, be unable to work out my very extensive investigations and possibly this will not happen within my lifetime since I am dreading the screams of the Boeotians if I were to make ideas public. (Letter from Gauss to Bessel [107], 1829)

Historical mathematical documents tend to be good source of unusual insults; the same is true here: Boeotia was a region of ancient Greece around the city Thebes whose inhabitants (the Boeotians) were considered dull by the Athenians [36]. The letters of Gauss, in general, make for interesting reading.

> I am surprised that you believe philosophers cannot be confused when it comes to notions and definitions. Nothing is more common, especially when it comes to philosophers who were not also mathematicians [...] Just have a look at our contemporary philosophers, look at Schelling, Hegel, Nees von Esenbeck and their friends, don't their definitions make your hair stand up? [...] Even Kant is not much better; his distinction between analytic and synthetic statements is, as I see it, a distinction that is either trivial or false. (Letter from Gauss to Schumacher [108], 1844)

5.2 We don't need any rules!

When people first started to use numbers to count things, they were not terribly concerned with axioms. Euclid famously used axioms, however, one could argue that the impact on mathematics itself has been limited: a triangle will be a triangle even if I have never heard of Euclid's axioms. The Pythagorean theorem was discovered before Euclid. Mathematicians used to have a rather relaxed understanding of the rules governing their game, they were trusting 'their gut'. If you ask people to think of a stereotypical mathematician, then 'easy-going, relaxed, not at all obsessed with details' is hardly what will come to mind. So how is it possible that mathematicians were playing fast and loose with the rules for several *thousand* years? At least partially, the philosopher Plato is to blame. Plato, a student of Socrates and the teacher of Aristotle, occupies a central position in Western philosophy. 'Plato' is actually a *nickname*, his birth name seems to have been Aristocles.

> And he learnt gymnastic exercises under the wrestler Ariston of Argos. And it was by him that he had the name of Plato given to him instead of his original name, on account of his robust figure, as he had previously been called Aristocles, after the name of his grandfather, as Alexander informs us in his Successions. But some say that he derived this name from the breadth of his eloquence, or else because he was very wide across the forehead, as Neanthes affirms. (Diogenes Laertius, [160])

Plato is the namesake of Platonic idealism. Since this is not a textbook of philosophy, we will be brief and inaccurate: in Platonic idealism, "ideas" are supposed to be the non-physical essence of all things, they exist forever and outside of time and never change. Everything that happens in life is merely an imitation of the perfect abstraction. If there are two cups of coffee in front of me, the 'two' is merely a shadow of the abstract idea of 'two' (or 'twoness') which exists forever and unchangeably so. Whenever I draw a circle, it is bound to be imperfect but there exists the idea of an abstract 'perfect' circle that is distinct from us and lives in a different realm. Moreover, good news, the soul can access these perfect forms.

> **Theaetetus:** I assign them to the class of notions which the soul grasps by itself directly. (Plato, Theaetetus, 186a [228])

Plato was ready to defend this notion.

> When Plato was discoursing about his 'ideas', and using the nouns 'tableness' and 'cupness': *I, o Plato!*, interrupted Diogenes, *see a table and a cup, but I see no tableness or cupness.* Plato answered, *That is natural enough, for you have eyes, by which a cup and a table are contemplated; but you have not intellect, by which tableness and cupness are seen.* ([160])

One way or another, most mathematicians are Platonic idealists: they believe that the number π truly exists as an abstract entity that has existed for all time and will continue to exist for all time, an abstract identity that we access with the mind. The logician Kurt Gödel (more about him in §19) says this quite literally

> But, despite their remoteness from sense experience, we do have something like a perception also of the objects of set theory, as is seen from the fact that the axioms force themselves upon us as being true. I don't see any reason why we should have less confidence in this kind of perception, i.e., in mathematical intuition, than in sense perception. (Gödel [111])

The author once asked a colleague whether they thought of themselves as Platonist. 'What do you mean?', the colleague asked. 'Well, do you believe that π exists forever and independently of humans?', the author asked. 'What a stupid question', came the reply, 'of course'. Delighted by this exchange, the author has since asked many other mathematicians and always received the same reply (the single exception being a Professor at the University of Virginia who claimed that mathematics is not that different from a really funny game that we invented and are now playing). Returning to why it is all Plato's fault: if π is something that truly exists outside of ourselves and something we go to with our mind when thinking about it, then we do not need to spend much time defining it. π is the abstract idea our soul goes to when we think about the area of the unit disk!

Distraction: Plato's Number. It is somewhat amusing that Plato himself, concerned with the idea of a mind accessing the ideal truth, has also given a description of a number that must go down in history as the most confusing description of a number of all time (a slightly too concise transcription might be to blame [192]). The text can be found in Plato's *Republic*, a recent translation starts out enigmatically

> Now for divine begettings there is a period comprehended by a perfect number, and for mortal
> by the first in which augmentations dominating and dominated when they have attained to
> three distances and four limits of the assimilating and the dissimilating, the waxing and the
> waning [...] (Plato, The Republic, translation by P. Shorey [229])

It goes on like this, *a basal four-thirds wedded to the pempad yields two harmonies at the third augmentation...* – and has confused people for 2000 years. The Roman philosopher and politician Cicero writes in a letter to his friend Atticus in 49BC

> I didn't guess your riddle: it is more obscure than Plato's number. (Cicero to Atticus, [56])

Various values for what Plato's number *might* have been have been proposed, among these are 216, 1728, 3600, 12960000. We refer to Dupuis [74] for more details and more proposed numbers. It stands to reason that Plato's number is an abstract idea that we will not be able to access using only our soul.

5.3 Everything explodes: the Barber paradox

At this point, we are able to describe what turned out to be a massive problem for the 'relaxed' approach to mathematics. Proceeding as a happy Platonist would, we can think of a set as a collection of 'objects' with a certain property. The set of all nonnegative integers

$$\mathbb{N} = \{0, 1, 2, 3, 4, \dots\}$$

would be such a collection. The curly brackets $\{\}$ indicate that they enclose a certain collection of things that we consider ('a set'); the order of elements in the set does not matter. One can think of it as a big bucket containing things that are of interest to us. Another useful piece of notation is $x \in A$, which simply means that x is part of the set A. For example $3 \in \mathbb{N}$ simply means '3 is contained in the set of integers' or, more concisely, '3 is an integer'. If one wanted to appear very scientific, then instead of saying 'π is not an integer', we could also write $\pi \notin \mathbb{N}$. Being done with notation, we now return to the set of all nonnegative integers $\mathbb{N}$. We know that 17 is in the set and we know that π is not. Starting with $\mathbb{N}$, one can build more complicated sets: for example

$$A_1 = \{\text{all integers that are multiples of } 3\}.$$

It is clear that the number 6 is in A but the number 7 is not. To make things more amusing, now that we are thinking of sets as abstract ideas that we access with our soul (as a Platonist would say), we may think of collections of sets. For example,

$$A_2 = \{\text{all sets that contain the integer } 3\}.$$

This set A_2 is now a pretty big set. It contains, for example, the set of positive integers $\mathbb{N}$ (since 3 is a positive integer) and it also contains the finite set $\{2, 3, 4\}$ and the set A_1. It also contains the set $\{3, \text{apple}, \text{banana}, \text{carrot}\}$. There are certain sets it does not contain. It does not contain the set of prime numbers that are bigger than 7, the set $\mathbb{P}_{>7} = \{11, 13, 17, 19, \dots\}$ and it also does not contain the set of all cars that were built between 1923 and 1924. There are more complicated sets: for example, if we consider the set of all symbols that can be produced on a computer

$$\left\{ \emptyset, \Delta, \int, \partial, \frac{\overline{\xi}}{\zeta}, \oplus^{\oplus}, \frac{\Phi\Xi}{\aleph\beth}, \prod_{\daleth}^{\Psi} \approx \cdots \right\},$$

then this set would be indeed in A_2 because '3' is a symbol that can be produced on this computer. One could worry a little whether the symbol '3' on the computer is really the same as the number 3, that is a good point and for this we refer to our discussion of signifier and signified in §2.5. The Barber paradox will rear its ugly head when we define the set B, where

> B contains exactly those sets that do *not* contain themselves.

This is a complicated set and we start with some examples. For example, B contains the set of nonnegative integers $\mathbb{N} = \{0, 1, 2, \dots, \}$ because $\mathbb{N}$ contains only individual numbers, it does not contain any sets and it certainly does not contain itself. Conversely, 'the set of all sets that were ever considered by a human being' is certainly a set that contains itself since I am a human being who just considered it. The problem now arises from a very simple question *Does the set B contain itself?* Well, it either does or it does not. If it does not contain itself, then B would be one of those sets that do not contain themselves (much like $\mathbb{N}$). But B is the collection of all sets with that property, so then B would contain itself. But if B contains itself, if it were one of those sets, then it would not be in B because B is exactly the collection of sets that do not contain itself. We are faced with the following dilemma:

> If B does not contain itself, then it does contain itself.
> If B contains itself, then it does not contain itself.

B has to either contain itself or not but either scenario leads to a self-defeating contradiction! So it seems that the set B cannot exist or maybe it exists but we are not allowed to think about it or maybe no sets exists or .... – suddenly the entire idea of mathematical objects as objects that exist separately from us and that we access with our soul seems a little more complicated than it did in the beginning. Is it possible for our soul to be misled?

5.4 Where's the barber?

Russell's paradox has a nice story version involving a barber. Imagine a town with a male barber who shaves exactly those men in the town who do not shave themselves. So we are defining

$$A = \{\text{men in the town who do not shave themselves}\}$$

and that is the set of people shaved by the barber. Does the barber shave himself? If the barber does not shave himself, then – since the barber shaves all the men in the town who do not shave themselves and he is a man in the town not shaving himself – he would shave himself. So what is the problem if he does? The problem is that the barber, by definition, only shaves those men who do not shave themselves – but he is a man who *does* shave himself and therefore he cannot shave himself. Either case presents a problem. The story is well suited to illustrate why this paradox had a lot of people worried: when we are asked to imagine a town in which a male barber lives, well, there is no difficulty in imagining it, there is the platonic idea of towns in which a male barber lives. Unless one is extraordinarily gifted in logical reasoning, there is no immediate mental resistance to the picture: colloquially, this is how one would describe a barber. However, as we have just seen, such a mental image cannot represent an abstract idea that is self-consistent. So how do we know that our idea of the number '8' is actually self-consistent? Or maybe, and that is the problem, maybe the number 8 is just as much a fiction as our idea of the town with that barber?

5.5 Barber paradox: history and consequences

The Barber paradox uses an idea that appeared thousands of years earlier under the name *liar paradox* by the Greek philosopher Epimenides who discusses the inhabitants of the Greek Island Crete.

Epimenides was a Cretan who made the statement: "All Cretans are always lying."

As is said in the Bible, *there is nothing new under the sun,* and just to prove that point, the Liar paradox itself can be found in the Bible (though, it seems, not to emphasize the logical paradox)

> One of themselves, even a prophet of their own, said, The Cretans are always liars, evil
> beasts, slow bellies. This witness is true. Wherefore rebuke them sharply. (Paul to Titus,
> 1:12, [28])

The liar paradox does not quite work as stated. If the statement were true, then Epimenides would be a Cretan who at some point told the truth and the statement would be false. But the statement could be false: the statement being false would mean that there is at least one Cretan who at least at one point in time told the truth: it need not be Epimenides, it may have been Orestis, the barber, who, 254 years ago, told the truth about liking the carrot soup he was eating. A proper formulation of the paradox would be the following (going back to Eubulides of Miletus)

> A man says that he is lying. Is what he says true or false?

By now we are quite familiar with this structure: if what he says is the truth, then he would be lying, in which case he would not be telling the truth. If he was lying about lying, then he would be truthful and we are again in trouble. We know very little about Eubulides. Diogenes Laertius [160] says that

> Eubulides of Miletus, who handed down a great many arguments in dialectics; such as the
> Lying one; the Concealed one; the Electra; the Veiled one; the Sorites; the Horned one; the
> Bald one. And one of the Comic poets speaks of him in the following terms
>
> > *Eubulides, that most contentious sophist,*
> > *Asking his horned quibbles, and perplexing*
> > *The natives with his false arrogant speeches*

Not at all of Eubulides' work had the same foundational impact on mathematics. His *Horns* paradox ('horned quibbles') goes as follows: 'That which you have not lost, you have. But you have not lost horns. Therefore, you have horns.' Diogenes of Sinope was not impressed by this argument: *A man once proved to him syllogistically that he had horns, so he put his hand to his forehead and said, 'I do not see them'* [160]. Greek philosophers had a special delight for these sort of word games and many exist: *if you say anything, what you say comes out of your mouth; but you say "a waggon," therefore a waggon comes out of your mouth* [160]. A more immediate consequence of the Barber paradox can be seen in the work of Gottlob Frege (1848–1925), a German philosopher, logician and mathematician. He was about to finish the second volume of his massive work *The Basic Laws of Arithmetic* when informed about the paradox by Bertrand Russell. Without worrying too much about the details, we may enjoy seeing the familiar structure in a letter from Russell to Frege

> I have encountered a difficulty only on one point. You assert ([Begriffsschrift] p. 17) that a
> function can also constitute the indefinite element. This is what I used to believe, but this
> view now seems to me to be dubious because of the following contradiction. Let w be the
> predicate: to be a predicate that cannot be predicated of itself. Can w be predicated of itself?
> (Letter from Russell to Frege, June 16, 1902 [90])

In the end, Frege realized that the paradox made the theoretical structure underlying his work untenable and he added the following note to his work.

> Hardly anything more unwelcome can befall a scientific writer than that one of the foun-
> dations of his edifice should be shaken, after the work is finished. I have been put in this
> position by a letter from Mr. Bertrand Russell. (Frege, *The Basic Laws of Arithmetic* [91])

Frege's willingness to admit an error is remarkable and this honesty had a special impact on Bertrand Russell who, more than 60 (!) years later, recalled the incident in a letter to the logician Jean van Heijenoort.

> As I think about acts of integrity and grace, I realise that there is nothing in my knowledge to compare with Frege's dedication to truth. His entire life's work was on the verge of completion, much of his work had been ignored to the benefit of men infinitely less capable, his second volume [i.e. of the Basic Laws of Arithmetic] was about to be published, and upon finding that his fundamental assumption was in error, he responded with intellectual pleasure clearly submerging any feelings of personal disappointment. It was almost superhuman.... (Letter from Russell to van Heijenoort, November 23rd, 1962. [128])

As a side remark, it is worth mentioning *Berry's paradox*, which is another nice illustration of the dangers that loom large in the foundations of mathematics. It seems to have first been published by Russell in 1908 who writes *This contradiction was suggested to me by Mr. G. G. Berry of the Bodleian Library* [249]. The idea is perfectly simple: the English language is capable of describing every natural number. Since there are infinitely many natural numbers, some of these numbers will require a *long* description in English where length is measured in syllables. In particular, this allows us to think of the 'smallest integer that requires at least 19 syllables' and Russell points out that this is 111777 since

one-hun-dred-and-e-le-ven-thou-sand-se-ven-hun-dred-and-se-ven-ty-se-ven

has 19 syllables. This all seems fairly innocent. The problem is shocking and surprising and we leave the full description to Bertrand Russell.

> Hence the name of some integers must consist of at least nineteen syllables, and among these there must be a least. Hence "the least integer not nameable in fewer than nineteen syllables" must denote a definite integer; in fact, it denotes 111,777. But "the least integer not nameable in fewer than nineteen syllables" is itself a name consisting of eighteen syllables; hence the least integer not nameable in fewer than nineteen syllables can be named in eighteen syllables, which is a contradiction. (Bertrand Russell, [249])

All these riddles and puzzles suggest that maybe *some* rigor is required. We need at least a little bit of precision and maybe having some fundamental rules, axioms, is not all that bad. To make that notion concrete, we discuss an axiomatic system for the natural numbers $\mathbb{N} = \{0, 1, 2, \dots\}$ (arguably an object that we certainly feel we understand very well). The Peano axioms are widely considered a good way of formalizing the foundations for the arithmetic of natural numbers. They are named after the Italian mathematician Giuseppe Peano (1858–1932) who invented many things, including a curve that fills all of space (see exercises) and *Latino sine flexione*, a simplified version of Latin. The Peano axioms are four rules.

1. *Zero axiom.* There exists a natural number, usually denoted as 0, which is the first and smallest natural number.
2. *Successor axiom.* Every natural number has a unique successor, which is also a natural number. This successor function is denoted as S, where $S(n)$ is the successor of the number n.

3. *No two successors are equal.* The successor of a number is never equal to any earlier number's successor. $S(m) = S(n)$ implies $m = n$.
4. *Induction axiom.* If a property holds for zero, and holds for the successor of any natural number whenever it holds for that number, then the property holds for all natural numbers.

Summarizing, the first rule states that the integers start somewhere and we call that number 0. For each integer n, there exists another integer $n + 1$ which succeeds it and, conversely, if we know that the number $n + 1$ is the successor of another integer, we can immediately deduce that that number had to be n. The last axiom deserves its own extensive discussion (see §5.7).

5.6 What is a proof?

A proof is the fundamental currency of mathematics: it is the reason *why* something is true. There are many different ways of defining the notion of proof, we will somewhat informally describe it as

> *a reason* why *a mathematical statement is true that is so compelling and so detailed that even your worst enemy, assuming they are also bound to follow the rules of logic and basic reason, could not help but agree that it is correct.*

It is easiest to start with an example.

1. Axiom: All cars are red.
2. If this object in front of me is a car,
3. then the object in front of me is red.

This is a fairly typical example: it would be difficult to disagree. Someone might argue that not all cars are red but the statement is not about that: it is a statement under the axiomatic assumption that indeed all cars are red. If we were to operate in a universe in which all cars are red, the argument is valid. Another person might point out that the object in front of me is not a car but a bicycle. In that case, also, nothing is wrong: the statement is only made under the assumption that the object in front of me is a car – if it happens to be a bicycle, nothing is asserted. This type of reasoning is usually ascribed to Aristotle ('Aristotelian logic') who first investigated it. Proofs in mathematics are best understood by looking at a large number of examples. It will then become clear what type of structure they tend to have and what properties they have in common. The act of *proving* a mathematical statement, essentially finding out why it has to be true and documenting the reasons in an unambiguous way, requires a lot of practice. Students of mathematics usually need several years to become acquainted with the techniques. Luckily, listening to music is easier than making music: it is the same with proofs and we will enjoy proofs without having to compose them.

Three types of proofs. Proofs, reasons why statements are true, 'usually' come in three different shapes. They tend to be either *direct* proofs, *indirect* proofs or *proofs by induction*. All three of them are valid ways of showing why something is true, the difference lies in their logical structure. The most concrete version is the *direct* proof: one starts with the assumptions and shows, step by step, that the assumptions imply the result. The argument above (all cars being red) is a representative example. *Indirect proofs* are more subtle. They employ the idea that a mathematical statement is either true or false. Accepting the fact that a mathematical statement is either true or false has a particularly nice consequence: each statement can be considered together with its negation

> '117 is a prime number' and '117 is *not* a prime number'

and for each such pair of statements exactly one of them is true and exactly one of them is false. This fundamental idea has the beautiful Latin name *tertium non datur* ('a third is not given'). This allows for an interesting approach: instead of trying to discover why a statement is true, we could instead try to discover why its negation is false. We will investigate this further in a subsequent section. For now, we will start by considering the third type, *proof by induction*.

Inductive reasoning. Before talking about *proofs by induction*, we discuss the (philosophical) problem of *inductive reasoning*. These two concepts, inductive reasoning and (mathematical) induction, are distinct but related. Both are fun. The process of inductive reasoning in philosophy refers to deducing a general law from a number of examples. Since we are only alive for a finite amount of time and can only experience a finite number of events, every time we guess a new natural law, for example, we do so based on a finite number of observations, we use inductive reasoning. However, and this is where it gets tricky, inductive reasoning can fail!

1. Every day that I have observed so far, the sun has risen.
2. Therefore, the sun will rise every day for all time.

This feels a little bit like trickery – and it is. We cannot help but remark that a more extreme argument would be possible: every day that I exist, the sun has risen. A hypothesis consistent with my observations so far is that the sun will rise exactly on those days that I exist. Or, as phrased by the philosopher Ludwig Wittgenstein,

> One could ask: may someone have telling grounds for believing that the earth has only existed for a short time, say since his own birth? Suppose he had always been told that, would he have any good reason to doubt it? Men have believed that they could make it rain; why should not a king be brought up in the belief that the world began with him? (Wittgenstein, *On Certainty*, §92, [302])

The problem is obvious: we cannot investigate an exhaustive list of infinitely many scenarios, that would take too much time. If we try to generalize from a finite number of observations, we may be misled. This dilemma was already summarized in ancient times, where it was argued that induction is not a reliable way to obtain certainty.

> It is also easy, I consider, to set aside the method of induction. For, when they propose to establish the universal from the particulars by means of induction, they will effect this by

> a review either of all or of some of the particular instances. But if they review some, the
> induction will be insecure [...] while if they are to review all, they will be toiling at the
> impossible [...] (Sextus Empiricus [261])

Sextus Empiricus was well trained in the art of doubt, he was a follower of Pyrrho of
Ellis, who may have been the ultimate skeptic. Pyrrho, a Greek philosopher, argued
that we were unable to know anything. Any type of knowledge we seem to have is
merely tradition or custom. He was willing to put his philosophy into practice: even
stepping off a cliff may not be bad and only appear to be so out of tradition – luckily,
he was saved by his friends.

> Owing to which circumstance, he (Pyrrho) seems to have taken a noble line in philosophy,
> introducing the doctrine of incomprehensibility, and of the necessity of suspending one's
> judgment, [...] For he used to say that nothing was honorable, or disgraceful, or just, or
> unjust. And on the same principle he asserted that there was no such thing as downright
> truth; but that men did everything in consequence of custom and law. For that nothing was
> any more this than that. And his life corresponded to his principles; for he never shunned
> anything, and never guarded against anything; encountering everything, even waggons for
> instance, and precipices, and dogs, and everything of that sort; committing nothing whatever
> to his senses. So that he used to be saved, as Antigonus the Carystian tells us, by his friends
> who accompanied him. [...] And he lived to nearly ninety years of age. (Laertius [160])

There is a simple reason why inductive reasoning has been troubling philosophers for
more than two millennia: there is no reason why it should be true (and it frequently
fails) but somehow we are forced to keep using it. Most of us have acquired a lot
of experience with gravity: things (including our bodies) tend to stick to the floor.
However, while we have accumulated a lot of experience it is not inconceivable that
we might wake up tomorrow to find that the law of gravity has been suspended. It
does not appear to be likely and it does not seem to have happened before but we do
not have the type of certainty that one has in mathematics.

5.7 Proof by (Mathematical) Induction

After all the trouble with inductive reasoning discussed in the preceding section, we
can now return to the calming effect of mathematical certainty by discussing the idea
behind *proof by induction*. We have already seen it hinted above when studying the
Peano axioms. Peano's fourth axiom is

> *Induction axiom.* If a property holds for zero, and holds for the successor of any natural
> number whenever it holds for that number, then the property holds for all natural numbers.

Instead of trying to find out why a certain statement might be true for all natural
numbers n, one could instead try to find out two separate things: (a) why it is true
for $n = 0$ and (b) why, *if* it is true for n, it must also be true for the successor $n + 1$.
This sounds a bit abstract and will be illustrated by the next result.

Theorem. *For every positive integer n, we have*

$$n \leq 2^n.$$

We can first try to figure out some special cases: clearly, when $n = 1$, we have $1 \leq 2$ which is true. When $n = 2$, we have $2 \leq 4$, when $n = 3$ we have $3 \leq 8$, for $n = 4$ we have $4 \leq 16$ and the statement seems clearly true. Moreover, the function 2^n appears to grow much more rapidly. While this may be a good heuristic or even convincing supporting evidence, it is not yet a rigorous mathematical argument by itself.

Proof by Induction. Instead of trying to argue for some arbitrary integer n that $n \leq 2^n$, we use the principle of Induction to break the problem into two separate arguments: a specific special case, say $n = 0$, and the induction step. The specific case is easy, we already did that. The Induction step consists of showing that *if the statement is true for one number, it must also be true for the successor.* More formally, we may assume that the statement is true for some integer n, so $n \leq 2^n$ and we have to show that this automatically implies that it will also be true for the successor, meaning $n + 1 \leq 2^{n+1}$. The reason this is useful is because we are now allowed to use the inequality $n \leq 2^n$. One way of arguing could now be

$$n + 1 \leq n + n = 2 \cdot n \underbrace{\quad \leq \quad}_{\text{here we use the assumption}} 2 \cdot 2^n = 2^{n+1}. \qquad \square$$

It is not entirely clear where this idea first arose. A good candidate [288] is the Sicilian mathematician, astronomer and priest Francesco Maurolico (1494–1575). Maurolico was interested in everything. In his *First Book of the Sphere* he suggested that Copernicus should be whipped for proposing the Earth revolves around the Sun. He also worked on music theory, wrote a history of Sicily and tried to explain the rainbow. Maurolico writes in his *Arithmeticorum Libri Duo* (Second Book of Arithmetic) that he has a new idea

> Therefore, we will attempt to explain those numerical forms that occur beyond this, demonstrating many things in an easier way and supplementing what other authorities either neglected or did not notice. (Maurolico [190])

His new idea is exactly Induction! Maurolico illustrates his idea on many examples, one of them being particularly nice: observe, he remarks,

$$1 = 1^2 \qquad 1 + 3 = 2^2 \qquad 1 + 3 + 5 = 3^2 \qquad 1 + 3 + 5 + 7 = 4^2.$$

Theorem. *The sum of the first n odd integers is a square number*

$$1 + 3 + \cdots + (2n - 1) = n^2.$$

Proof. We prove this result using Induction. It is true for $n = 1$ and, also by the examples above, for $n = 2, 3, 4$. Suppose now it is true for a fixed value n. Then we may write the case for $n + 1$ as

$$1 + 3 + \cdots + (2n - 1) + (2n + 1) = (1 + 3 + \cdots + (2n - 1)) + (2n + 1)$$
$$= n^2 + (2n + 1) = (n + 1)^2. \qquad \square$$

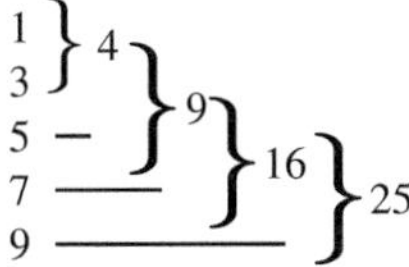

Fig. 5.2: The diagram appearing Francesco Maurolico's 1575 *Two Books of Arithmetic*: *From the aggregation of odd numbers taken successively from unity in order, square numbers are constructed...* [190]

This beautiful fact has been rediscovered a great many times. The Russian mathematician Andrei Kolmogorov recalls that *The first impression of mathematical 'discovery' I experienced at the age of five or six, noticing the regularity in the identities* $1 = 1^2, 1 + 3 = 2^2, 1 + 3 + 5 = 3^2, 1 + 3 + 5 + 7 = 4^2$ *and so on* [279]. It is even featured in a great science fiction story!

> As a child of seven, while investigating the house of a relative, Renee had been spellbound at discovering the perfect squares in the smooth marble tiles of the floor. A single one, two rows of two, three rows of three, four rows of four: the tiles fit together in a square. Of course. No matter which side you looked at it from, it came out the same. And more than that, each square was bigger than the last by an odd number of tiles. It was an epiphany. (Ted Chiang, *Division by Zero* [55])

Sometimes a proof by induction can be almost like a method. We illustrate this with a particularly nice argument that goes back to Solomon Golomb (1932–2016). It may help to keep in mind that Golomb is sometimes credited with creating the mathematical theory behind the game *Tetris*. The following result is truly a Tetris result. We quickly introduce two pieces generalizing the familiar domino piece.

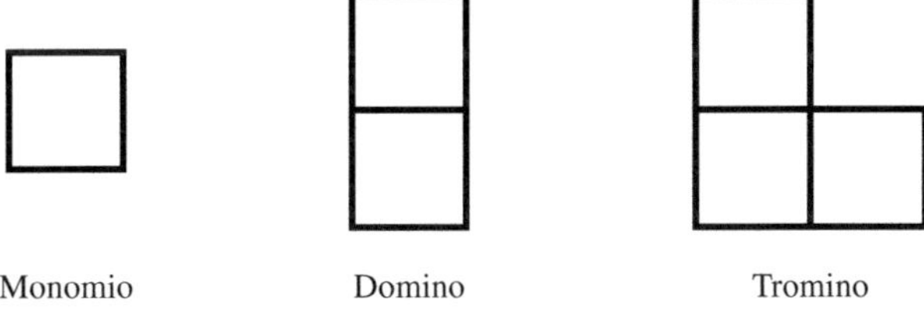

Fig. 5.3: Un (mono), dos, tres: a single tile, the domino tile and a tromino tile.

Can we cover the usual 8×8 chessboard with monomios? The answer is obviously yes, we will need 64 monomios, one for each square. It is almost as easy to see that we can also cover it with 32 dominos. Finally, we observe that we *cannot* cover it with trominos because each tromino is using exactly 3 squares but 64 is not divisible by 3. In fact, $64 = 21 \cdot 3 + 1$, so there will always be at least one square left over. Golomb showed that this is the only obstruction: the moment one square is removed, any square, the rest can always be covered by trominos.

Theorem (Golomb). *Consider an 8×8 chess board. If a monomio is placed on an arbitrary square, the remaining 63 squares can always be covered with 21 trominos.*

This Theorem is also very well suited to illustrate another mathematical principle: sometimes it is easier to prove a more general result. A more general result may help focus one's attention on that which really matters. We prove a more general result.

Theorem (Golomb [113]). *Consider a chess board with dimensions $2^n \times 2^n$ where $n \in \mathbb{N}$. If a single monomio is placed on an arbitrary square, the remaining $2^{2n} - 1$ squares can always be covered with trominos.*

Proof by Induction. We start by investigating the case $n = 1$, corresponding to a 2×2 chess board. This case is pretty easy: no matter which square is removed, the shape that is left over is exactly a (possibly rotated) tromino. As is often the case, the most interesting part of a proof by induction is to see how the case n implies the case $n + 1$. For simplicity, we illustrate $1 \to 2$. Suppose we are given a 4×4 chess board and a monomio has been removed.

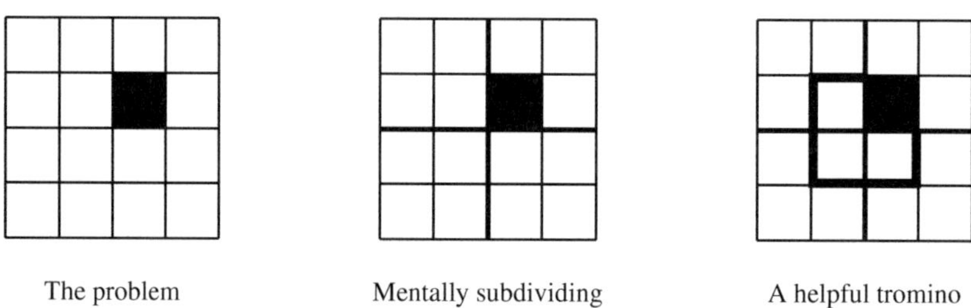

The problem	Mentally subdividing	A helpful tromino

More generally, given a $2^{n+1} \times 2^{n+1}$ chess board with a single square missing, we can think of it as four $2^n \times 2^n$ chess boards, one of which has a single square missing. We can then add a tromino so that the other 3 also have exactly one square missing. However, that is an easier problem that, by induction, we already know how to solve and this completes the argument. $\qquad\qquad\square$

The trick is as follows: we may, mentally, think of the 4×4 chess board as four 2×2 chess boards placed next to each other. One of these four chess boards has a missing square. We can now place a single tromino right in the center so that the other four 2×2 squares also have a missing square. This leads us to a situation where we have four 2×2 chess boards each of which has exactly one square missing: but this problem we already solved. This proof is more than a reason *why* the statement is true, it actually encodes a way of solving the problem: indeed, the argument could be implemented on a computer! The result, involving only tetris pieces and chess boards, also manages to tell us something new about integers! A statement that follows relatively quickly from another statement is called a 'Corollary'. Here is our first one.

Corollary. *The number $4^n - 1$ is divisible by 3.*

Two proofs. A $2^n \times 2^n$ chess board has 4^n squares. After removing 1, we are left with $4^n - 1$ squares that can be tiled with trominos and must therefore be a multiple of 3. This may seem like a really complicated proof and we can give a second reason why the statement is true: we use induction. Checking small values of n, we get $4^1 - 1 = 3$ and $4^2 - 1 = 15$ and $4^3 - 1 = 63$ and all these numbers can clearly be divided by 3. One can now write

$$4^{n+1} - 1 = 4 \cdot (4^n - 1) + 3.$$

If $4^n - 1$ is divisible by 3, then so is $4 \cdot (4^n - 1)$. The number 3 is clearly divisible by 3 and the sum of numbers divisible by 3 remains divisible by 3. $\qquad \square$

5.8 Induction, Ulam, rabbits and horses

Recall that the Ulam sequence starts $a_1 = 1, a_2 = 2$. After that a_n is the smallest positive integer that can be uniquely written as the sum of two distinct earlier terms.

Theorem. *The n-th Ulam number a_n satisfies*

$$a_n \leq 2^{n-1}.$$

Proof. It is certainly true in the beginning since $a_1 = 1 \leq 2^0$ and $a_2 = 2 \leq 2^1$. Suppose now it's true for the first n Ulam numbers. We will show that it is also true for the $(n + 1)$-th Ulam number. The argument is simple: if there are no Ulam numbers between a_n and $a_{n-1} + a_n - 1$, then $a_{n-1} + a_n$ can be uniquely written as the sum of two distinct earlier terms and it is the smallest number with that property. This shows that $a_{n+1} \leq a_n + a_{n-1}$. Such an inequality is perfect for induction since we now have $a_{n+1} \leq a_n + a_{n-1} \leq 2^{n-1} + 2^{n-2} \leq 2^{n-1} + 2^{n-1} \leq 2^n$. $\qquad \square$

What happens if we just keep using the inequality $a_{n+1} \leq a_n + a_{n-1}$? This leads to the celebrated FIBONACCI SEQUENCE named after the Italian mathematician Leonardo da Pisa (1170–1240), who was also known as *Leonardo filius Bonacci*, Fibonacci.

> A certain man had one pair of rabbits together in a certain enclosed place, and one wishes to know how many are created from the pair in one year when it is the nature of them in a single month to bear another pair, and in the second month those born to bear also. (Fibonacci, *Liber Abaci* [262])

Biology aside, this leads to an elementary arithmetic problem, the answer to which is given by a sequence where each term is the sum of the two previous terms

$$1, 1, 2, 3, 5, 8, 13, 21, 34, 55, \ldots$$

Formally, the Fibonacci sequence is defined by $F_0 = 1$, $F_1 = 1$ and then, for $n \geq 2$, by $F_n = F_{n-1} + F_{n-2}$. This is maybe the most famous integer sequence in all of mathematics. It is so famous that there is an entire scientific journal, the *Fibonacci Quarterly*, dedicated to its study: this journal has existed since 1963 and is still being printed. At this point, we observe the following inequality.

Theorem. *The n-th Ulam number a_n is smaller than the n-th Fibonacci number*

$$a_n \leq F_n.$$

Proof. This is again a proof by induction. We note that $a_1 = 1 = F_1$ and $a_2 = 2 = F_2$, so the inequality is certainly true in the beginning. Then, assuming the inequality holds true for $a_1, \ldots, a_n$, we argue that $a_{n+1} \leq a_n + a_{n-1}$, which is the inequality we already used above. Using the induction assumption, we have $a_n \leq F_n$ and $a_{n-1} \leq F_{n-1}$ and thus $a_n + a_{n-1} \leq F_n + F_{n-1}$. Finally, using the definition of the Fibonacci numbers $F_n + F_{n-1} = F_{n+1}$. Combining all these inequalities, we end up with $a_{n+1} \leq F_{n+1}$, as required. $\square$

After Fibonacci's rabbits, it seems appropriate to move on to a famous result of Pólya concerning horses. It is famous because it is wrong: your task is to find the mistake.

Theorem (Pólya). *All horses have the same color.*

Proof by Induction (find the mistake). If there is a single horse, then all horses have the same color. Let us now assume that in a group of $n + 1$ horses, it is true that each group of n horses is comprised of horses that all have the same color. By excluding one horse at a time and applying the assumption, all horses share the same color. $\square$

This should not be confused with a paradox from Chinese philosophy.

Theorem (Gongsun Long, 320–250 BC). *A white horse is not a horse.*

> Is "a white horse is not horse" assertible?
> Advocate: It is.
> Objector: How?
> Advocate: "Horse" is that by means of which one names the shape. "White" is that by means of which one names the color. What names the color is not what names the shape. Hence, one may say "white horse is not horse."

5.9 Exercises

1. Contemplate the ontological status of 'the circle' or π. Has π always existed? Or has it been constructed at some point in the last 5000 years? If humanity goes extinct, is π going to stick around?
2. A rather extreme case involving axioms is the following: in 1821, the French mathematician Cauchy made a mathematical claim that is now considered to be false. However, starting in the 1960s, there have been ongoing discussions about whether the claim is false or whether Cauchy was working in an axiomatic framework that is different from what we use today [58, 66, 151]. Find the last mistake you made and see whether you can blame it on a different set of axioms.
3. Try to reformulate the Barber paradox in an entirely new story that encapsulates the same paradox but avoids Barbers in a village.

4. Think about Berry's paradox. Think about it for so long that you feel you can explain it to a friend. Try to explain it to a friend. Then think about it some more.
5. Get a chess board (or draw one), remove one of the 64 squares and actually carry out Golomb's argument. This will make the proof really clear!
6. Proofs by Induction get easier with practice. Use Induction to prove that

$$\underbrace{1 + 1 + 1 + \cdots + 1}_{n \text{ times}} = n.$$

7. Use Induction to prove that

$$1 + 2 + \cdots + n = \frac{n(n+1)}{2}.$$

There is another proof that was, legend has it, discovered by the mathematician Carl Friedrich Gauss when he was a very young boy. Instead of computing $S = 1 + 2 + \cdots + n$, we can compute twice the sum, reorder the terms and match them up in the correct way

$$
\begin{aligned}
2S &= (1 + 2 + 3 + \cdots + n) + (1 + 2 + 3 + \cdots + (n-1) + n) \\
&= (1 + 2 + 3 + \cdots + n) + (n + (n-1) + (n-2) + \ldots 2 + 1) \\
&= (1 + n) + (n + 1) + \cdots + (n + 1) = n(n + 1).
\end{aligned}
$$

See whether you can work out the details behind the argument.
8. Giuseppe Peano is also famous for discovering the first *space-filling curve*. In an 1890 paper entitled *On a Curve that Fills an Entire Plane Area* [225] he describes the *arc of curve that passes through all the points of a square*. A related but slightly simpler construction was given by David Hilbert and looks as follows. Stare at the picture long enough to figure out what is happening and then draw the $n = 3$ case from memory (it might help to first draw $n = 1$ and $n = 2$).

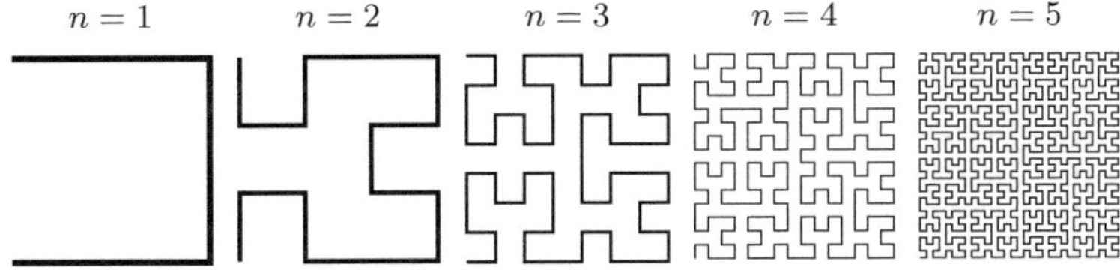

9. Have a look at Peano's 1903 article describing *De Latino Sine Flexione* [226]. *Lingua latina fuit internationalis in omni scientia, ab imperio Romano, usque ad finem saeculi XVIII [...] Sed non tota lingua latina est necessaria.* If you enjoy this, have a look at some other artificially constructed languages, they are fun!
10. Use Induction to prove that

$$1 + 2^2 + 3^2 + \cdots + n^2 = \frac{n(n+1)(2n+1)}{6}.$$

11. A truly spectacular and very beautiful result is described by Nicomachus of
    Gerasa in his *Introduction to Arithmetic*, where he points out that

$$1^3 + 2^3 = (1 + 2)^2$$
$$1^3 + 2^3 + 3^3 = (1 + 2 + 3)^2$$
$$1^3 + 2^3 + 3^3 + 4^3 = (1 + 2 + 3 + 4)^2.$$

Use Induction to prove the Theorem of Nicomachus. Nicomachus was a dedicated
Pythagorean, so dedicated that some books even list it as part of his name.

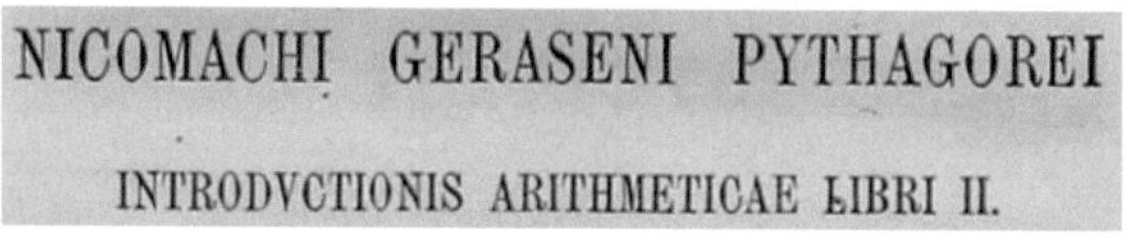

NICOMACHI GERASENI PYTHAGOREI

INTRODVCTIONIS ARITHMETICAE LIBRI II.

Fig. 5.4: *Nicomachus of Gerasa, the Pythagorean* [215].

12. Admire

Chapter 6
Direct and Indirect Proofs

> We cannot pretend to offer proofs. *Proof* is an idol
> before whom the pure mathematician tortures himself.
> (Arthur Eddington, *The Nature of the Physical World* [78])

6.1 Direct Proof

A proof is a *reason* why something is true. There can be many reasons why something is true and often there are many proofs. It is sometimes nice to have different reasons why something is true. Proofs tend to come, usually, in three different flavors.

1. **Direct.** You just keep arguing until you get there.
2. **Indirect.** You assume the opposite and show that you run into trouble.
3. **Induction.** We already saw this above.

Direct proofs are exactly what one would expect. The argument starts from reasonable premises and finishes when reaching the desired result. Here is an example.

Theorem. *If n is an even integer, then n^2 is also an even integer.*

Direct proof. Let n be an even integer; this means that the integer is divisible by 2 or, equivalently, that dividing by 2 yields another integer. Formally, if n is even then this means there exists an integer m such that $n = 2m$. Consider the square of n

$$n^2 = (2m)^2 = 4m^2.$$

Since $4m^2$ is of the form $2 \times (2m^2)$, where $2m^2$ is an integer, we can express n^2 as $2 \times \ell$, where $\ell = 2m^2$. Therefore, n^2 is also an even integer. In fact, the argument shows the stronger result that n^2 is even divisible by 4. $\square$

This is somehow not too surprising, it is an argument that begins with what we know, makes some inferences, draws some conclusions and ends up at the right place. This does not mean that direct proofs are always simple, some direct proofs are tremendously complicated. However, since the purpose of this section is to survey *methods* of proofs rather than proofs themselves, we will move to the next and much more interesting way of proving things: the method of indirect proof, which is also known as proof by contradiction.

© The Author(s), under exclusive license to Springer Nature Switzerland AG 2025
S. Steinerberger, *The Unreasonable Elegance of Mathematics*,
Springer Undergraduate Mathematics Series, https://doi.org/10.1007/978-3-032-03815-9_6

6.2 Indirect proof

Indirect proofs are based on the fundamental principle that if we are given a statement ('The sun is shining') and its negation ('The sun is not shining'), then exactly one of the two statements is true and exactly one of the two statements is false. One can quickly come up with settings where this is not the case and that is usually indicative of a lack of precision in what is meant ('Is the sun shining somewhere or shining on me?'). There is a very famous example of an indirect proof. It goes back to the Pythagoreans, the group of people centered around Pythagoras. Pythagoras is well known for $a^2 + b^2 = c^2$, however, this was already known to the Babylonians. It is less well-known that he was a bit of a cult leader. One of the principles of the cult was that beans are strictly off-limits.

> He also forbade his disciples to eat beans, because, as they were flatulent, they greatly partook of animal properties [he also said that men kept their stomachs in better order by avoiding them]; and that such abstinence made the visions which appear in one's sleep gentle and free from agitation. (Diogenes Laertius, [160])

Other reasons have been brought forward, some of them most curious

> And Aristotle says, in his treatise on Beans, that Pythagoras enjoined his disciples to abstain from beans, either because they resemble some part of the human body, or because they are like the gates of hell (for they are the only plants without parts); or because they dry up other plants, or because they are representatives of universal nature, or because they are used in elections in oligarchical governments. He also forbade his disciples to pick up what fell from the table, for the sake of accustoming them not to eat immoderately, or else because such things belong to the dead. (Diogenes Laertius, [160])

Beans ended up playing a big role in the life of Pythagoras. He appears to have died in the ancient city of Crotona, now in Italy, at the hands of the locals: and, once more, beans are involved.

> [...] some say that the people of Crotona themselves did this, being afraid lest he might aspire to the tyranny. And that Pythagoras was caught as he was trying to escape; and coming to a place full of beans, he stopped there, saying that it was better to be caught than to trample on the beans, and better to be slain than to speak; and so he was murdered by those who were pursuing him. (Diogenes Laertius, [160])

The Pythagorean school had many other beliefs. One such belief involves the idea that all numbers can be expressed as the ratio of integers, $1/2, 2/3, 14/25, \ldots$. Discovering that this is not true, that there exist some numbers that cannot be written in this way, so-called *irrational numbers*, is also usually attributed to the Pythagorean school. Since it violated a most sacred principle, punishment was swift.

> Indeed the sect (or school) of Pythagoras was so affected by its reverence for these things that a saying became current in it, namely, that he who first disclosed the knowledge of [...] irrationals [...] perished by drowning (Commentary of Pappus, [220])

We will now study this famous proof. Recall that $\sqrt{2} \sim 1.4142\ldots$ is the unique positive number with the property that multiplying it by itself results in 2. This number shows up prominently in geometry as the length of the hypotenuse of one of the simplest triangles. It is thus clear that the number exists since we can identify it as the length in a triangle. The question is whether it is a rational number (for example, $577/408$ is quite close). Note that there are hypotenuses of right-angled triangles that are definitely rational, for example $\sqrt{5^2 + 12^2} = 13$ is a rational number.

Theorem. *$\sqrt{2}$ is irrational, it cannot be written as the ratio of two integers.*

Proof. This is *the* classical indirect proof, also known as a proof by contradiction. Assume, for the sake of contradiction, that $\sqrt{2}$ is rational. This means that $\sqrt{2}$ can be expressed as the fraction a/b, where a and b are integers. We can even assume that one of the two numbers a, b is odd: if they are both even, then we could divide both of them by 2 since

$$\frac{a}{b} = \frac{a/2}{b/2}$$

and we could continue this until either the numerator or the denominator is odd. We may assume that $\sqrt{2} = a/b$ with either a or b being odd. Squaring both sides

$$2 = \frac{a^2}{b^2} \qquad \text{which can be rewritten as} \qquad 2b^2 = a^2.$$

This equation shows that a^2 is an even number (because it can be divided by 2). However, the square of a number a^2 is even if and only if a is even (see also above). This means that a is even which forces b to be odd (since one of the two has to be odd). Since a is even, we can write $a = 2c$ for some integer c. Substituting this back into the equation,

$$2b^2 = (2c)^2 \quad \text{and thus} \quad 2b^2 = 4c^2 \quad \text{and therefore} \quad b^2 = 2c^2.$$

This shows that b^2 is even, which implies that b is also even, a contradiction. $\qquad\square$

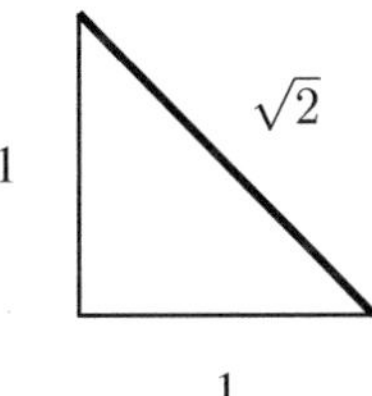

The irrationality of $\sqrt{2}$ has a great number of consequences. The most unexpected is perhaps the one brought forth by Roger Bacon (1219–1292), a Franciscan friar who was so admired that his nickname was *Doctor Mirabilis*. In his major work, literally called *Opus Majus* ('Greater Work') [17], Bacon uses the irrationality of $\sqrt{2}$ to argue against the existence of atoms!

> [...] it does not follow [...] that the world is composed of an infinite number of material particles called atoms, as Democritus and Leucippus maintained, by whose position Aristotle and all students of nature have been more hindered than by any other error. Yet this error is wholly eliminated by the power of geometry; for no stronger argument can be used

against this error than that the diagonal of the square in that case and its side would be commensurable, that is, would have a common measure [...] the contrary to which Aristotle already teaches. (Roger Bacon, *Opus Majus* [17])

Spelled out, the argument is as follows: if the side of square would be a certain number of atoms, n, and if the diagonal is made up of m atoms, then

$$\frac{m}{n} = \sqrt{2} \qquad \text{which is a contradiction}$$

and therefore atoms do not exist. This is a bold argument, and we will leave it to the reader to figure out whether there is a problem in the argument or whether atoms truly do not exist. Roger Bacon should not be underestimated, in the 1627 *The Famous Historie of Fryer Bacon: Containing the wonderfull things that he did in his Life* [18] it is reported that Bacon had a powerful device

> It is spoken of before now, that Fryer Bacon had a Glasse, which is of that excellent nature, that any man might behold any thing that he desired to see within the compasse of fifty miles round about him: With this Glasse he had pleasured divers kinds of people: for Fathers did oftentimes desire to see [...] how their Children did [...] (*Famous Historie of Fryer Bacon*)

The same manuscript also describes how Friar Bacon and Friar Bungey, *a great Scholler and a Magician (but not to bee compared to Fryer Bacon)*, made a brass head and enlisted the Devil to make it talk

> To do this they prepared all things ready and went one Evening to a Wood thereby, and after many ceremonies used, they spake the words of coniuration, which the Devill straight obeyed and appeared unto them, asking what they would? Know, said Fryer Bacon, that wee have made an artificiall head of Brasse, which we would have to speake, to the furtherance of which wee have raised thee, and being raised, we will keepe thee, unless thou tell to us the way and manner how to make this Head to speake. (*Famous Historie of Fryer Bacon*)

Fig. 6.1: Bacon's *artificiall head of Brasse* coming to life [21].

The adventure continues and we will leave it to the reader to see how it ends. The story is also retold in a comedy written by Robert Greene around 1590, *The Honorable Historie of Frier Bacon and Frier Bongay*. Towards the end of his life, we are told, Bacon abandoned his wizardy: *His time hee spent in Prayer, Meditations, and such Divine Exercises, and did seeke by all means to perswade men from the study of Magicke. Thus lived he some two years space in that cell, never comming forth: his meat and drink he received in a window, and at that window he did discourse with those that came to him; His grave he digged with his own nayles, and was laid there when dyed. Thus was the Life and Death of this famous Fryer, who lived most part of his life a Magician, and dyed a true penitent Sinner, and an Anchorite.* [18]

6.3 Computing $\sqrt{2}$

Now that we know that $\sqrt{2}$ is not a rational number, it is even more tempting to better understand its elusive nature. This has been a persistent problem for a long time, an ancient Indian collection of texts (the *Sulbasutras*) describes it as

> The measure is to be increased by its third and this [third] again by its own fourth less the thirty-fourth part [of that fourth]; this is [the value of] the diagonal of a square ([234])

which corresponds to

$$1 + \frac{1}{3} + \frac{1}{3 \times 4} - \frac{1}{3 \times 4 \times 34} = \frac{577}{408} = 1.41421\ldots$$

which is a remarkably good approximation. We recall that if $\sqrt{2} = a/b$, then this would mean that $a^2 - 2b^2 = 0$ and, as we saw above, this is impossible. However,

$$577^2 - 2 \cdot 408^2 = 1$$

which is as close to 0 as a non-zero integer can be. This idea can be generalized into a remarkably powerful idea that leads to powerful approximations of $\sqrt{2}$ by means of the *Pell sequence*.

> The Englishman, John Pell (1611–1875), after whom the Pell equation is named, had essentially nothing to do with it. His name got attached to it due to his correspondence on the subject with Euler (1707–1783). This equation has been known since time immemorial. [...] According to André Weil [...], *a cat may not know how to define a rat, but when it smells one, it knows how to find it.* Almost a millennium and a half ago [...] Brahmagupta might never have heard of an 'infinite cyclic group,' but while studying the Pell equation, he knew he was dealing with one. (Chahal, [50])

The Pell sequence $0, 1, 2, 5, 12, 29, 79, 169, 408, 985, \ldots$ is defined by setting $P_0 = 0$, $P_1 = 1$ and, for all $n \geq 2$, the formula $P_n = 2P_{n-1} + P_{n-2}$. It is a bit similar to the Fibonacci sequence in the sense that the first two terms are prescribed and the subsequent terms follow as a combination of the two preceding ones (though now the previous one is multiplied by 2 first). Using a proof by induction, we can establish a concrete formula for how they behave.

Theorem. *We have*

$$P_n = \frac{(\sqrt{2}+1)^n - (1-\sqrt{2})^n}{2\sqrt{2}}.$$

Proof. When $n = 0$, then using $x^0 = 1$, we have

$$\frac{(\sqrt{2}+1)^0 - (1-\sqrt{2})^0}{2\sqrt{2}} = \frac{0}{2\sqrt{2}} = 0 = P_0,$$

which is correct. When $n = 1$, a simple computation shows that the formula gives $1 = P_1$. If it is now correct up to P_n, then

$$P_{n+1} = 2P_n + P_{n-1} = \text{some computation} = \frac{(\sqrt{2}+1)^{n+1} - (1-\sqrt{2})^{n+1}}{2\sqrt{2}}. \quad \square$$

So far, so good – but what does this have to do with $\sqrt{2}$ and approximating it by rational numbers? Recall that if $\sqrt{2} = a/b$, then $2b^2 - a^2 = 0$. We know that this equation cannot have any integer solutions $a, b \in \mathbb{N}$ because $\sqrt{2}$ is irrational. Moreover, if a and b are integers, then $2b^2$ and a^2 are also integers and $2b^2 - a^2$ is also an integer – since it is an integer that is different from 0, the closest it can be to 0 is ± 1. This is exactly what happens with the Pell sequence!

Theorem. *We have*

$$(P_{n-1} + P_n)^2 - 2P_n^2 = (-1)^n.$$

Proof. Plug in the above formula and simplify. $\quad \square$

This formula can also be rewritten as

$$\left(\frac{P_{n-1} + P_n}{P_n}\right)^2 - 2 = \frac{(-1)^n}{P_n^2},$$

which shows that $(P_{n-1} + P_n)/P_n$ are excellent approximations to $\sqrt{2}$. Using this formula and looking back at the Pell numbers $0, 1, 2, 5, 12, 29, 79, 169, 408, 985, \ldots$, note that 408 is familiar from the *Sulbasutras* that contained the approximation 577/408. It is now clear what 577 is, it is simply $408 + 169$. More generally, the Pell numbers give rise to increasingly good approximations to $\sqrt{2}$,

$$\frac{3}{2}, \frac{7}{5}, \frac{17}{12}, \frac{41}{29}, \frac{108}{79}, \frac{248}{169}, \frac{577}{408}, \ldots$$

Heron's method. Pell numbers are a wonderful way to approximate $\sqrt{2}$, but there is a certain magic ingredient to them. It is not immediately clear how one would change things to compute $\sqrt{3}$. There is a much more geometric method that goes back to the mathematician Heron of Alexandria (first century AD). The idea is beautifully simple: $\sqrt{2}$ is the side-length of a square with area 2. So how about we start with an initial guess for $\sqrt{2}$? One such guess could be $x = 1$. We can then construct a rectangle with area 2 having one side-length be x. This is easy: the other side-length has to be $2/x$. If we have a rectangle with area 2, then one of the side-lengths is at most $\sqrt{2}$ and another is at least $\sqrt{2}$: if they were both larger than $\sqrt{2}$, then their product would be larger than 2. If they were both smaller than $\sqrt{2}$, then their product would be smaller than 2. Heron's idea is now simple: the average should be a better estimate. Repeating the procedure, if we have a rectangle with sides x_n and $2/x_n$ and we hope that their average will be a better estimate, we arrive at the sequence

$$x_{n+1} = \frac{x_n + 2/x_n}{2} = \frac{x_n}{2} + \frac{1}{x_n}.$$

Let us see how well this idea works in practice: we start with $x_1 = 1$. Then

$$x_2 = \frac{1}{2} + \frac{1}{1} = \frac{3}{2} = 1.5.$$

When we plug that approximation back in, we get

$$x_3 = \frac{3/2}{2} + \frac{1}{3/2} = \frac{3}{4} + \frac{2}{3} = \frac{17}{12} \sim 1.4166\ldots$$

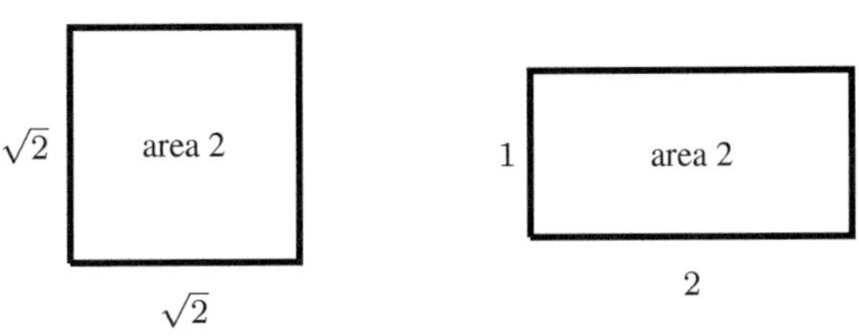

Fig. 6.2: For a rectangle with area 2, one side-length is $\leq \sqrt{2}$ and the other is $\geq \sqrt{2}$.

We plug it back in one more time and get

$$x_4 = \frac{17/12}{2} + \frac{1}{17/12} = \frac{17}{24} + \frac{12}{17} = \frac{577}{408} = 1.4142\ldots$$

That one is familiar! That is the approximation coming from the *Sulbasutras*! Heron's method, motivated by geometric considerations, actually recovers the Pell numbers. Also, it can immediately be generalized: suppose we try to approximate $\sqrt{a}$ where $a > 0$ is some positive number. If we have an approximation $x_n \sim \sqrt{a}$, then the two numbers x_n and a/x_n are the side-lengths of a rectangle with area a. One of the two numbers has to be at most $\sqrt{a}$ and one is at least $\sqrt{a}$. Therefore, their average is hopefully closer and we may hope that

$$x_{n+1} = \frac{x_n + a/x_n}{2} = \frac{x_n}{2} + \frac{a}{2x_n}$$

is a better approximation! We can try this out for $\sqrt{3} = 1.73205\ldots$. Starting with $x_1 = 1$, we get $x_2 = 2/1 = 2$ which, plugging back in, gives us $x_3 = 7/4 = 1.75$, which, plugging back in, gives us $x_4 = 97/56 = 1.7321\ldots$ which is a good approximation. Also,

$$2^2 - 3 \cdot 1^2 = 4 - 3 = 1$$
$$7^2 - 3 \cdot 4^2 = 49 - 3 \cdot 16 = 1$$
$$97^2 - 3 \cdot 56^2 = 9409 - 3 \cdot 3136 = 1.$$

This is no coincidence! There is a nice explanation for all of these miracles: it requires a little bit more mathematics than we will go into here, but the curious reader is suggested to read more about the *continued fraction expansion*.

6.4 Exercises

1. Prove that $\sqrt{3}$ is irrational.
2. Figure out whether Roger Bacon's argument has a problem, maybe a missing assumption? Or does the irrationality of $\sqrt{2}$ really show that atoms do not exist?
3. Adapt the $\sqrt{2}$ irrationality proof to show that $\sqrt{4}$ is irrational: this is false, some part of the proof needs to fail. Can you identify which part that is?
4. If we have a right-angled triangle with sides a, b, c, then the Pythagorean theorem tells us that $a^2 + b^2 = c^2$. If all a, b, c are integers, then a and b have to be distinct numbers: if they were the same, $a = b$, then $2a^2 = c^2$, which would imply that $\sqrt{2}$ is rational, which is false. However, we can show that there are infinitely many triangles where a and b are integers with difference 1. Using P_n to denote again the Pell sequence, show that

$$a = 2P_n P_{n+1} \qquad b = P_{n+1}^2 - P_n^2 \qquad c = P_{n+1}^2 + P_n^2$$

leads to a choice of parameters that work. You may want to use

$$P_{n+1}P_{n-1} - P_n^2 = (-1)^n.$$

5. A Pythagorean triple (a, b, c) is a collection of three integers such that $a^2 + b^2 = c^2$. Examples are $3^2 + 4^2 = 5^2$ and $5^2 + 12^2 = 13^2$. Find some more, find a pattern, and use the pattern to prove that there are infinitely many.
6. Use Heron's method to approximate $\sqrt{17}$. A good starting guess is $x_1 = 4$ since $4^2 = 16$ and 16 is close 17.
7. Heron suggests that if $x_1 > 0$ is a good approximation for $\sqrt{a}$, then

$$x_2 = \frac{x_1 + a/x_1}{2}$$

is an even better approximation. Prove that $x_2 \geq \sqrt{a}$ and then $x_n \geq \sqrt{a}$ for all $n \geq 3$. Heron's method always approximates the square root from above.

Chapter 7
Dirichlet's Box Principle

This section is dedicated to a funny idea that may be considered a particular type of direct proof. The basic form is illustrated in Figure 7.1: if we are given three balls and asked to put them into 2 containers in any way that we like, then there is going to be a container that has at least two balls in it. This is not too surprising, if each container only had one ball or less than there would still be a ball left over. More generally, if we are given $n + 1$ balls and are asked to distribute them in n containers, at least one container will have two balls.

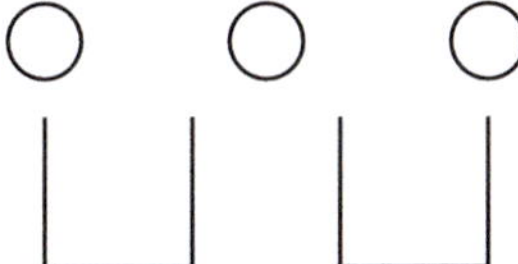

Fig. 7.1: A very scientific diagram illustrating the box principle: if we put 3 balls into 2 containers then there is at least one container containing at least 2 balls.

7.1 The number of hairs

The Dirichlet principle was used more than 200 years before Dirichlet by the French Jesuit priest Jean Leurechon (1591–1670). Leurechon published a book with a title that roughly translates to *The most beautiful selected statements from all of mathematics*, [171].

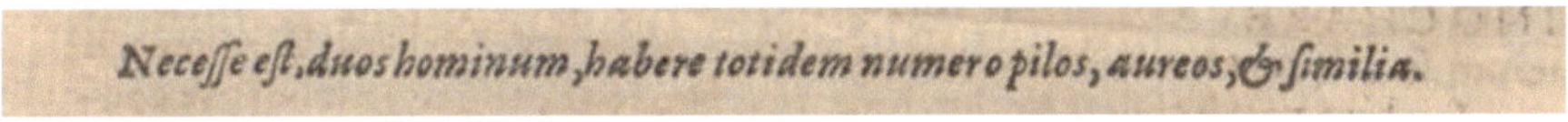

Necesse est, duos hominum, habere totidem numero pilos, aureos, & similia.

Fig. 7.2: Leurechon [171] writing down the box principle *Necesse est, duos hominum, habere totidem numero pilos, aureos, & similia* 'It is necessary that two men have the same number of hairs, gold and such things'.

Theorem. *There are two people with the same number of hairs on the head.*

Proof. We give two proofs. The first proof is very short: two bald people exist. The second proof is more along the lines of what Leurechon had in mind: there are more than a million people in the world. People have at most 150,000 hairs on their head. If we put it each person into a box (describing how many hairs they have on their head), then there are two people who end up in the same box. □

© The Author(s), under exclusive license to Springer Nature Switzerland AG 2025
S. Steinerberger, *The Unreasonable Elegance of Mathematics*,
Springer Undergraduate Mathematics Series, https://doi.org/10.1007/978-3-032-03815-9_7

Leurechon's collection of the most beautiful statements is a rich document. It begins by restating a Platonic conception of mathematics: mathematics is an abstract ideal that we access with our mind, it does not necessarily exist in reality: *The circle of geometry cannot be found on earth, in gold, in the heavens nor in any of the natural things* [171]. Not all of his statement are of such a philosophical nature: one is concerned with how many people could fit on the surface of Earth and he arrives at the number 148456800000000.

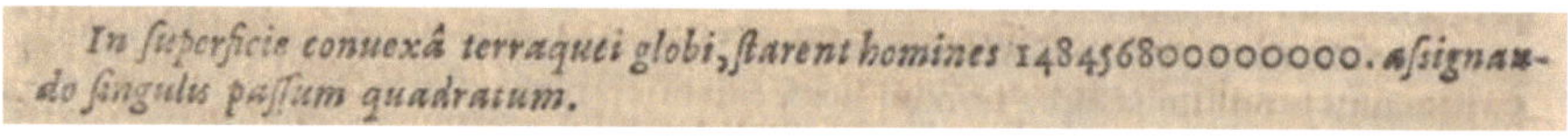

In superficie connexâ terraquei globi, starent homines 148456800000000. assignando singulis passum quadratum.

Fig. 7.3: Leurechon [171] computing *precisely* how many people fit on Earth.

7.2 Numbers close to each other

Real numbers are simply the things that we ordinarily think of as numbers ('all the numbers'). They include the set of all nonnegative integers, denoted by $\mathbb{N}$, but also negative integers like -8 (that set is usually written as $\mathbb{Z}$), they include all the rational numbers like $3/7$ (often denoted by $\mathbb{Q}$) and also irrational numbers like $\sqrt{2}$ or π. The real numbers are usually denoted by $\mathbb{R}$.

Theorem. *If $0 \leq x_1, x_2, \ldots, x_{11} \leq 1$ are any eleven real numbers in $[0, 1]$, then there exist two different numbers x_i and x_j such that*

$$|x_i - x_j| \leq \frac{1}{10}.$$

Proof. We consider the ten intervals $[0, 0.1]$ and $[0.1, 0.2]$ and so on until $[0.9, 1]$ as ten different boxes. Here $[a, b]$ is an abbreviation which means: 'all the numbers that are at least as big as a and not bigger than b'. Each of the eleven numbers is contained in at least one box. Since we have eleven numbers and only ten boxes, at least one box has to contain two numbers. However, each box only has length $1/10$, so any two numbers in the same box cannot have a distance that is larger than $1/10$. This simple result is even optimal in the sense that the constant $1/10$ cannot be replaced by any smaller number, which can be seen by considering the example $x_1 = 0, x_2 = 1/10, x_3 = 2/10$ all the way until $x_{11} = 1$. $\qquad\square$

A short diversion: normal numbers. Having introduced the set of real numbers $\mathbb{R}$, there is a fun problem worth discussing. Numbers can go on forever. This is even true for rational numbers, for example $1/3 = 0.3333\ldots$ Rational numbers are a little bit limited: their digit expansion is always going to be periodic, for example $1/11 = 0.09090\ldots$ and, more advanced, $1/13 = 0.076923076923076\ldots$ Real numbers can go on forever without repeating themselves. This led to a funny question: are there real numbers whose digit expansion has the property that all possible digit combinations appear roughly the same number of times in the long run? This would mean that the frequency with which we find '123' is the same as the frequency with which we find '231' or '777'. Such numbers are called *normal*. It is easy to give examples of numbers that are not *normal*. $1/3$ is such an example, the digit '3' appears more frequently than the digit '1', which never appears. A good question is whether *normal* numbers exist. This was answered by Émile Borel in 1909.

Theorem (Borel [32]). *'Most' real numbers are normal. In particular, if you pick a random number between 0 and 1, it is very likely to be normal.*

If most real numbers are normal, we should have no problem coming up with an example. Let's start with a concrete case: our favorite real number $\sqrt{2}$. Among its first million digits, the digit 1 appears 98.925 times, the digit 2 appears 100.436 times, the digit 3 appears 100.190 times and so on. It clearly seems to be the case that all digits appear with the same frequency. $\sqrt{2}$ seems to be normal, but this is actually still not proven and it seems like a very difficult problem.

Open problem. Is $\sqrt{2}$ normal?

So we find ourselves in a very strange situation: most real numbers are normal, but we struggle to come up with an example. Luckily, in the case of normal numbers, a very nice example was provided by Champernowne in 1933. The number is so simple that, after seeing it, one can never forget it: it is known as Champernowne's constant.

Theorem (Champernowne [51]). *The number*

$$0.1234567891011121314151617\ldots \qquad \textit{is normal.}$$

We still do not know a good way to tell whether any given number is normal: most of them are, but it is very difficult to find specific examples. Mathematicians sometimes refer to this as the problem of *finding hay in a haystack*: it is hard to believe that this can be a problem, but normal numbers are a good example.

7.3 Friends

In any given group of people, any two people may or may not be friends. We assume that friendship is symmetric, if A is friends with B then B is also friends with A. This is an entirely reasonable assumption, Aristotle *was once asked what a friend is; and his answer was, 'One soul abiding in two bodies.'* [160]. We represent people by dots and indicate friendship by drawing lines between people.

Fig. 7.4: Different friendship arrangements: 4 people that do not know each other, 2 pairs of friends that have never met, a group of 3 friends and a person they are not friends with, a group of 3 friends with one of them bringing a new friend along.

Many different such arrangements are possible, and some are shown in Fig. 7.4. It is easy to come up with many more examples, simply draw some points on a piece of paper and connect some of them by lines (the lines do not need to be straight lines, they simply indicate friendship). One way to analyze a given social situation could be by counting how many friends each person has within the group.

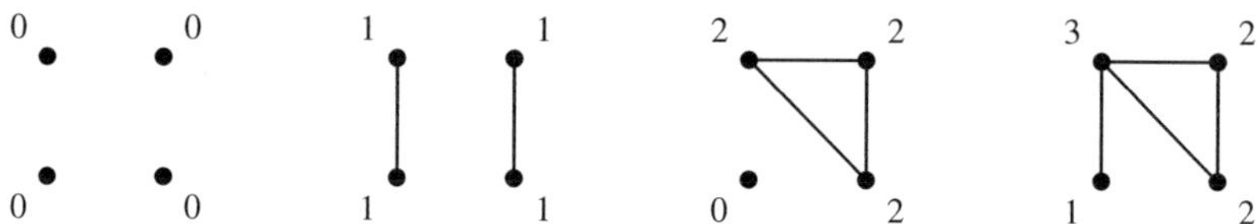

Fig. 7.5: Counting the number of friends.

If one plays the game long enough, see also Fig. 7.5, one observes that, independently of the social configuration, there are always at least two people who have exactly the same number of friends in that group. This is unavoidable!

Theorem. *In any group of people, there always at least two people who have the same number of friends in that group.*

Proof. Suppose the group has n people. Each person has a number of friends that is a number in the list $\{0, 1, 2, \ldots, n-1\}$. It will be helpful to consider two cases.
Case 1. Everybody has at least one friend. Let us suppose that everybody has at least one friend. Then we have n people and they all have a number of friends that is in $\{1, 2, \ldots, n-1\}$. But this corresponds to distributing n balls (people) into $n-1$ boxes (possible number of friends they have). The box principle applies, and we know that there have to be at least 2 people with the same number of friends.
Case 2. Someone has no friends. Suppose there exists a person who has no friends in that group, let's call the person P. This means that among the remaining $n-1$ people, nobody can have $n-1$ friends: any person with $n-1$ friends would have to be friends with everyone else, including P, but then P would have a friend. It is now possible that there is another person, say Q, that also has 0 friends. Well, then we would be done, because P and Q would be two different people with the same number (0) of friends. The only case that is remaining is therefore the case where P has 0 friends and everyone else has at least 1 friend. In that case, there are $n-1$ people whose number of friends is somewhere in $\{1, 2, \ldots, n-2\}$, and the box principle applies again! $\square$

Meeting people. Suppose we host a dinner party with $n \geq 28$ people. Conversation is not going well, time for some math magic! (Fair warning: it has been the author's experience that nothing is better at breaking up parties than a bit of math magic.) For each pair of guests A and B we define

the $A - B$ meeting birthday as the day of the year in which A and B were together in the same room at the same time for the very first time.

Since everybody is in the same room now, everybody has an $A-B$ birthday. However, especially if this is a group of old friends, some may have met under all sorts of crazy circumstances. People may have known each other for many years. However, if there are enough people, then there are always two $A-B$ birthdays that happened on the same day, like April 12 or July 22 (maybe not in the same year, though).

Theorem. *In any group of $n \geq 28$ people, there are at least two $A-B$ birthdays that happened on the same day of the year.*

Proof. We use the box principle. Given n people, we first count how many different pairs there are. Person 1 has $A-B$ birthdays with $(n-1)$ different people leading to the $A-B$ birthdays of $(1,2),(1,3),\ldots,(1,n)$. Person 2 has at some point met person 1, but we already counted that one above as $(1,2)$. The meetings person 2 has had that we have not yet considered are with the remaining $n-2$ people, the meetings being $(2,3),(2,4),\ldots,(2,n)$. This shows that there is a total of

$$(n-1)+(n-2)+\cdots+1 = \frac{(n-1)n}{2} \qquad A-B \text{ birthdays.}$$

This formula can be proven by induction, the answer is also $\binom{n}{2}$, the number of ways of picking 2 out of n people. If $n \geq 28$, then the number of pairs is larger than the number of days in a year since $(n-1)n/2$ is then at least $(27 \cdot 28)/2 = 378 \geq 366$. $\square$

7.4 Rational Approximations

One of the most spectacular applications of the box principle is the Dirichlet Approximation Theorem. It deals with a fundamental question: suppose we are given a number $x \in \mathbb{R}$, how well can we approximate x using rational numbers? There are two cases: either x is itself a rational number (or maybe even an integer) or it is irrational (like $\sqrt{2}$). In the first case, x being rational, there is no problem, we can write x as a rational number and we can do it in infinitely many different ways like

$$5 = \frac{25}{5} = \frac{100}{20} = \cdots \qquad \text{or} \qquad \frac{1}{7} = \frac{2}{14} = \frac{3}{21} = \cdots .$$

Suppose now that x cannot be written as a rational number. A good example, already familiar, is the irrational number $\sqrt{2} \sim 1.4142\ldots.$ It can never be written as a fraction but, as seen before, we can approximate it using fractions: how well can we hope to do this?

Question. How small can we make the error $|x - \frac{p}{q}|$? How does it depend on q?

Here, $|x-p/q|$ makes use of the 'absolute value' function $|x|$ which has the definition

$$|x| = \begin{cases} x & \text{if } x \geq 0 \\ -x & \text{if } x \leq 0. \end{cases}$$

Ignore the sign, we only care about the distance between x and 0, we do not care whether it is positive or negative. In our case, asking about $|x - p/q|$ simply means asking how close p/q can get to x, we do not care whether it is a little bit larger or a little bit smaller. There is a basic observation.

Theorem. *For each $x \in \mathbb{R}$ and each $q \in \mathbb{N}$, there exists $p \in \mathbb{N}$ such that*

$$\left| x - \frac{p}{q} \right| \leq \frac{1}{2q}.$$

Proof. The proof is direct: we show how to find such a p. If $x \approx p/q$, then, multiplying both sides by q, we have $qx \approx p$. This suggests a simple method:

1. Compute $q \cdot x$.
2. If $q \cdot x$ is an integer, then set p to be $p = q \cdot x$.
3. If $q \cdot x$ is not an integer, then let p be the integer closest to $q \cdot x$.

We note that if $q \cdot x$ is an integer, then $p = q \cdot x$ is an integer and $x = p/q$. There is no error at all, we achieve perfect approximation. If $q \cdot x$ is not an integer, how far away is the nearest integer going to be? Some examples quickly show that for any positive number $y > 0$, the nearest integer is at most distance $1/2$ away: either by rounding up or by rounding down, whatever is more efficient. This means that there exists $|r| \leq 1/2$ such that $p = q \cdot x + r \in \mathbb{N}$. Then $p/q = x + r/q$ and the result follows. $\square$

This is, in some sense, optimal. If someone gives us a number $x > 0$ to be approximated and a denominator q, we will usually not be able to do much better than that. However, maybe there are *some* values of q for which one can do better? To understand the efficiency of the method, we define the function $f(q)$ which is the error we are making when approximating $\sqrt{2}$ with a rational number of the form p/q

$$f(q) = \min_{p \in \mathbb{N}} \left| \sqrt{2} - \frac{p}{q} \right|.$$

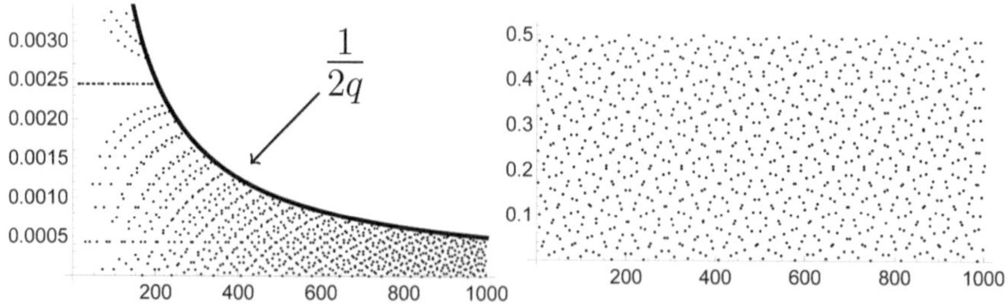

Fig. 7.6: Left: the values of $f(q)$ compared to $1/(2q)$. Right: the values of $q \cdot f(q)$.

The values of $f(q)$ for $1 \leq q \leq 10000$ are shown in Fig. 7.6. As guaranteed by the Theorem, we have $f(q) \leq 1/(2q)$. This can, alternatively, be written as $q \cdot f(q) \leq 1/2$, and both $f(q)$ as well as $q \cdot f(q)$ are shown in Fig. 7.6. The Theorem cannot be improved: sometimes $q \cdot f(q)$ is really quite close to $1/2$. However, we also see that sometimes the value of $q \cdot f(q)$ is much smaller than $1/2$, sometimes we get much better approximations. With this in mind, we return to the sequence of Pell numbers P_n given by $P_n = 2P_{n-1} + P_{n-2}$ starting

$$0, 1, 2, 5, 12, 29, 79, 169, 408, 985, \ldots$$

As observed before, this sequence has the wonderful property

$$\left(\frac{P_{n-1} + P_n}{P_n} \right)^2 = 2 + \frac{(-1)^n}{P_n^2}.$$

Taking a square root, we see (with a little bit of computation) that

$$\left| \frac{P_{n-1} + P_n}{P_n} - \sqrt{2} \right| \leq \frac{1}{2} \frac{1}{P_n^2}.$$

This is a *much* better approximation because

1. for every $x > 0$ and every $q \in \mathbb{N}$ there exists $p \in \mathbb{N}$ with

$$\left| x - \frac{p}{q} \right| \leq \frac{1}{2q}$$

2. for $x = \sqrt{2}$ there exists infinitely many $q \in \mathbb{N}$ (the numbers in the Pell sequence) for which there exists $p \in \mathbb{N}$ such that

$$\left| \sqrt{2} - \frac{p}{q} \right| \leq \frac{1}{2q^2}.$$

We note that the error in the second part, $1/q^2$, is much smaller than $1/q$: the price we have to pay is that this is not possible for every q but only for some very special q. The second part of the statement, these sporadic approximations that are much better, relies on the magical properties of the Pell sequence: maybe $\sqrt{2}$ is special here? The celebrated Dirichlet Approximation Theorem shows that this is not the case, there are always sporadic approximations that are *much* better. And what is even nicer is that the proof only needs the box principle, nothing else!

Theorem (Dirichlet Approximation Theorem). *For any real number x and any positive integer N, there exist integers p and q with $1 \leq q \leq N$ such that*

$$\left| x - \frac{p}{q} \right| < \frac{1}{Nq} \leq \frac{1}{q^2}.$$

The slightly more exciting way of stating the result is as follows.

Corollary. *For any real number x, there exist infinitely many integers (p, q) with*

$$\left| x - \frac{p}{q} \right| \leq \frac{1}{q^2}.$$

Proof. Let x be a real number, and let N be a positive integer. Consider the numbers

$$0, x, 2x, 3x, \ldots, Nx.$$

We do not care very much how large these numbers are, we only care about what they look like after the decimal point. Using $\lfloor x \rfloor$ to denote the largest integer that is $\leq x$, we will be interested in what is known as the *fractional part* $\{x\}$, defined by

$$\{x\} = x - \lfloor x \rfloor.$$

This may look complicated but is easy to illustrate: when considering an integer, there is nothing left after the decimal point, and we simply get $\{2\} = 2 - \lfloor 2 \rfloor = 0$. For a generic real number, we simply drop what is in front of the decimal point and keep that which comes afterwards. For example,

$$\left\{ \sqrt{2} \right\} = \{1.4142\ldots\}$$
$$= 1.4142\cdots - \lfloor 1.4142\ldots \rfloor = 1.4142\cdots - 1 = 0.4142\ldots$$

$\{x\}$ is always a number between 0 and 1

$$\{\pi\} = \pi - \lfloor \pi \rfloor = \pi - 3 = 0.1415\ldots$$

It is worth noting that this is another instance of ambiguity in notation: before, we saw $\{1, 2, 3, \ldots\}$ as a way of denoting sets, now $\{x\}$ is a function applied to a single number. Notation is context-dependent and it is arbitrary! We return to the $N + 1$ numbers $\{0x\}, \{x\}, \{2x\}, \{3x\}, \ldots, \{Nx\}$. Applying the box principle, we note that having $N + 1$ numbers in $[0, 1]$ means that there exist at least two such numbers that are distance $\leq 1/N$ from each other: let us call these numbers m and n. This implies that, for some integer A, $|m \cdot x - n \cdot x - A| \leq 1/N$. Then, however, using that $m \cdot x - n \cdot x = (m - n) \cdot x$ and dividing,

$$\left| x - \frac{A}{m - n} \right| \leq \frac{1}{|m - n|} \frac{1}{N}.$$

This is the desired result: set $p = A$ and $q = m - n$. $\qquad\qquad\square$

7.5 Converse of the Approximation Theorem

Every good result, such as the Dirichlet approximation theorem, begs the question: could an even stronger result be true? We now know there are infinitely many rational numbers such that $|x - p/q| < 1/q^2$, maybe there are also infinitely many rational numbers such that one has an even better approximation, maybe something like $|x - p/q| < 1/q^3$? As it turns out, this is not true: Dirichlet's theorem is really the best one can hope for in general. An example that illustrates this is again our old friend, the square root of 2.

Theorem. *For every rational number p/q we have*

$$\left| \sqrt{2} - \frac{p}{q} \right| \geq \frac{1}{10q^2}.$$

Proof. Instead of proving that $|\sqrt{2} - p/q| > 1/(10q)$, we are going to prove the equivalent statement that one obtains after multiplying by q

$$|q\sqrt{2} - p| > \frac{1}{10q}.$$

Here's the key insight: the number $2q^2 - p^2$ is an integer and we also know that it cannot be 0 (since otherwise $\sqrt{2}$ would be rational). This means that

$$|2q^2 - p^2| \geq 1.$$

Now we use the binomial $(a - b)(a + b) = a^2 - b^2$ to rewrite things

$$1 \leq |2q^2 - p^2| = |(q\sqrt{2} - p)(q\sqrt{2} + p)| = |q\sqrt{2} - p| \cdot |q\sqrt{2} + p|.$$

This is nice because it implies the inequality

$$|q\sqrt{2} - p| > \frac{1}{|q\sqrt{2} + p|}. \qquad (\diamond)$$

This inequality is almost what we want: the difficulty is that there is still a factor p in there, and we want to have an inequality that only involves q. There is a nice little trick: if the inequality that we want to prove were false, then p/q would have to be very close to $\sqrt{2}$, but then we know roughly how big p is: it would be approximately $\sqrt{2} \cdot q$. It remains to make this trick precise. We use a case distinction: either

$$|q\sqrt{2} - p| > \frac{1}{10q} \qquad \text{or} \qquad |q\sqrt{2} - p| \leq \frac{1}{10q}.$$

In the first case, we are already done, that is what we wanted to show. If the second inequality was true, then, dividing by q, we have

$$\left| \sqrt{2} - \frac{p}{q} \right| < \frac{1}{10q^2} \leq \frac{1}{10} \qquad \text{implying} \qquad \frac{p}{q} \leq \sqrt{2} + \frac{1}{10} \leq 2.$$

Therefore $p \leq 2q$ and $q\sqrt{2} + p \leq q\sqrt{2} + 2q \leq 4q$. In that case, we can use our inequality $(\diamond)$ and deduce that

$$\left| q\sqrt{2} - p \right| > \frac{1}{|q\sqrt{2} + p|} \geq \frac{1}{4q} \geq \frac{1}{10q},$$

which is exactly what we want. $\qquad\qquad\qquad\qquad\qquad\qquad\qquad\qquad\qquad\qquad\square$

This is a pretty fun result. One could wonder whether it is also true for other numbers, for example π. Is it true that for all $p, q \in \mathbb{N}$

$$\left| \pi - \frac{p}{q} \right| \geq \frac{1}{10^{100}} \frac{1}{q^2} \quad ?$$

This is a really hard question. This inequality is probably false, but we do not know for sure. We do know (this is implied by a result of Salikhov [251]) that there exists some constant $c > 0$ such that for all integers $p, q \in \mathbb{N}$, one has $|\pi - \frac{p}{q}| \geq c/q^8$, but obviously $1/q^8$ is much, much smaller than $1/q^2$.

7.6 Another $\sqrt{2}$ mystery

Our friend $\sqrt{2}$ has been a helpful guide. It suggested various results, it served as a motivation for other results, it was a nice example and just generally very, very helpful. It feels appropriate to finish with a little mystery. Suppose we want to write $\sqrt{2}$ as a computer would: in binary. Formally, this just means we write

$$\sqrt{2} = 1 + \frac{\mathbf{0}}{2} + \frac{\mathbf{1}}{4} + \frac{\mathbf{1}}{8} + \frac{\mathbf{0}}{16} + \frac{\mathbf{1}}{32} + \frac{\mathbf{0}}{64} + \frac{\mathbf{1}}{128} + \ldots$$

which one also sometimes abbreviates as

$$\sqrt{2} = (1.0110101\ldots)_2,$$

the result of writing the number $\sqrt{2}$ using only powers of 2, its binary expansion. A natural question is: do the digits 0 and 1 appear with roughly equal proportions? Among the first 100 digits, 0 appears 46 times and 1 appears 54 times. Among the first 1000 digits, 0 appears 494 times and 1 appears 506 times. So, yes, it does seem that they appear with the same frequency. Nobody knows how to prove this!

Theorem. *Among the first n binary digits of $\sqrt{2}$, the digit 1 has to appear at least $\sqrt{n}/2$ many times.*

Proof. The idea behind the argument is simple (checking all the details is a good exercise). We take the first n digits and suppose that k of them are equal to 1. Writing only the terms that contain the digit 1, we have

$$\sqrt{2} = 1 + \underbrace{\frac{1}{4} + \frac{1}{8} + \cdots + \frac{1}{2^n}}_{k \text{ terms}} + \text{rest}.$$

We will ignore the rest, this is a very small number anyway. If we then take the first few terms and square it out, we know what we have to get: we have to get a number that is incredibly close to $\sqrt{2}$, just a little bit smaller, and then its square has to be incredibly close to 2, just a little bit smaller, and thus

$$\left(1 + \frac{1}{4} + \frac{1}{8} + \cdots + \frac{1}{2^n}\right)^2 = 1.999999999\ldots999324\ldots$$

$$= 1 + \frac{1}{2} + \frac{1}{4} + \frac{1}{8} + \cdots + \frac{1}{2^{n-5}} + \text{rest}.$$

If the sum on the left has only k terms, then squaring it out can only produce k^2 terms, but we need $k^2 \geq n/2$ and this gives the desired result. $\qquad\square$

Of course, one would expect that among the first n digits of the binary expansion of $\sqrt{2}$ both digits 0 and 1 appear approximately $\sim n/2$ times. The best known lower bound currently seems to be $\sim \sqrt{2n}$ due to Vandehey [289], see also [20].

7.7 Exercises

1. A more general form of the box principle is as follows: if we distribute n items over m boxes, then there is at least one box containing $\lfloor n/m \rfloor$ items. Here $\lfloor x \rfloor$ denotes the smallest integer that is at least as large as x. In particular, if x is an integer itself, then $\lfloor x \rfloor = x$. Try to understand why that is.
2. Suppose we have n points in a square with side-length 1. Show that there exist two points whose distance is $\leq 1000/\sqrt{n}$. (The number 1000 does not need to be this large but it may simplify some of your arguments).
3. Can you construct examples of friendship diagrams of $5, 6, 7, \ldots$ people so that *exactly* two people have the same number of friends (and not more)?
4. If we have $n \geq 28$ people, we are always going to find 2 people with the same 'meeting' birthday. How many people are needed to ensure that we can find 3 people with the same 'meeting' birthday?
5. We have seen that, for general real numbers x, finding approximations like $|x - p/q| \leq 1/q^2$ is the best we can hope for. However, there are some real numbers where one can get better results. Find a real number such that there are infinitely many rational numbers p/q with $0 < |x - p/q| \leq 1/q^3$. Hint: try things like

$$x = 1 + \frac{1}{10} + \frac{1}{10^{10}} + \frac{1}{10^{10^{10}}} + \cdots.$$

6. Show that if a real number x has the property that there exist infinitely many rational numbers p/q with $0 < |x - p/q| < 1/q^2$, then x is irrational. Hint: suppose $x = a/b$ is irrational. Then

$$0 < \left| x - \frac{p}{q} \right| = \left| \frac{a}{b} - \frac{p}{q} \right| = \frac{|aq - bp|}{bq} \le \frac{1}{q^2}.$$

   Since the expression is larger than 0, we have $|aq - bp| > 0$. This is an integer, therefore $|aq - bp| \ge 1$. Then, however, ...

7. Combine the last two problems to show that

$$x = 1 + \frac{1}{10} + \frac{1}{10^{10}} + \frac{1}{10^{10^{10}}} + \cdots$$

   is irrational. We do not know what it is, but surely it is not a rational number!

8. Define the sequence of positive $u_1 = 1$ and $u_{n+1} = \left\lfloor \sqrt{2}(u_n + 1/2) \right\rfloor$. Ron Graham and Henry Pollak [115] observed that $u_{2n+1} - 2u_{2n-1}$ is the n-th binary digit of $\sqrt{2}$. Surely one can use this insight to prove that many of these digits are 1? Warning, this may be difficult, the author does not know how to solve this problem.

Chapter 8
Adding Infinitely Many Things

8.1 Our friend Aristotle

Can we add infinitely many things and get something finite? This question is rarely asked in this form, but the underlying principle has occupied a lot of people for a long time. An example can be found in the work of the Greek philosopher Aristotle, who stated that

> That which is in locomotion must arrive at the half-way stage before it arrives at the goal. (Aristotle, *Physics* [9])

Before deciphering what this means (in §8.2), let us first put *Physics* in proper context. Aristotle's *Physics* is an attempt to understand the natural laws of the Universe. Figuring out how the Universe works is not an easy task and not everything in the works of Aristotle has survived the transition to the present day (this should not be read as a criticism of Aristotle, who managed to figure out a remarkable number of things, see Feyerabend [82]). Many of his observations remain correct to this day. A famous example that turned out to be wrong is that heavier objects fall faster.

> We see the same weight or body moving faster than another [...], other things being equal, the moving body differs from the other owing to excess of weight or of lightness. (Aristotle, *Physics* [9])

This was believed for a long time (Aristotle was quite the authority). Galileo (more about him in §11) gave a beautiful argument showing that the statement has to be wrong because it is not self-consistent.

Theorem. *Objects fall at the same rate.*

Proof by Galileo. Galileo argues as follows: suppose heavier things fall faster. We can take a rock weighing 1 pound and another rock weighing 2 pounds. The rock weighing 2 pounds falls faster than the rock weighing 1 pound because it is heavier. So if we now take these rocks and glue them together, the arising construction should fall slower than the rock weighing 2 pounds (because it is slowed down by the rock weighing 1 pound which falls at a slower rate). However, it could also be considered a rock weighing 3 pounds and should fall faster and that is a contradiction. $\square$

S. Steinerberger, *The Unreasonable Elegance of Mathematics*,
Springer Undergraduate Mathematics Series, https://doi.org/10.1007/978-3-032-03815-9_8

8.2 John and the cat

We can now unravel what Aristotle means when he says *That which is in locomotion must arrive at the half-way stage before it arrives at the goal*. The main idea goes back to the Greek philosopher Zeno, who proposed a curious thought experiment. John is chasing after his cat which has escaped. The cat is only interested in going for a nice stroll. John is worried and runs twice as fast as the cat. We assume that, initially, the cat has a nice headstart of one mile.

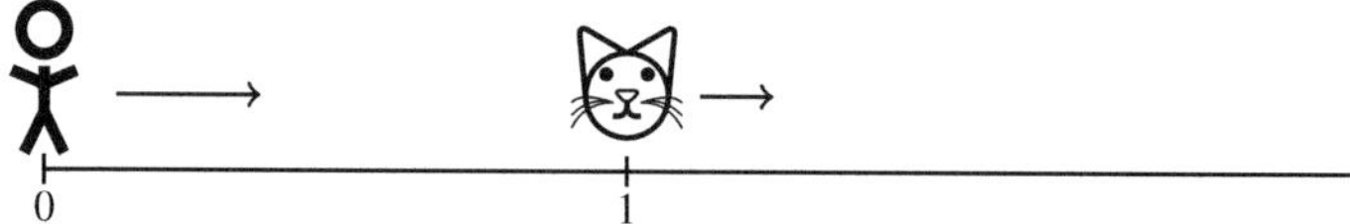

Without doing too much math, we imagine John running after his cat. At some point, he will have run for 1 mile and will reach the point where the cat originally was. In that time, however, the cat also moved: since it moves half as fast as John, it moved half a mile. The new situation therefore looks a little bit like this.

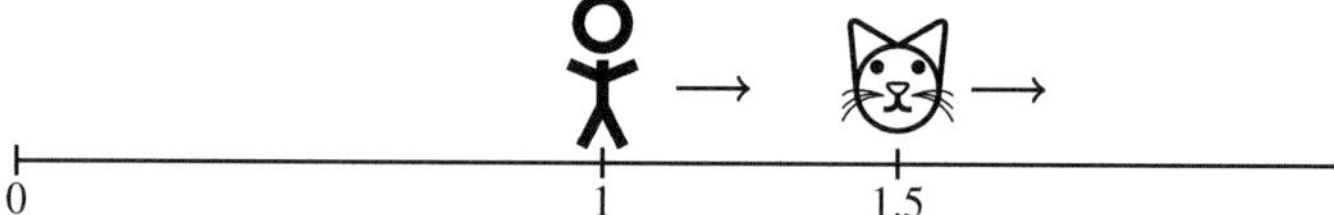

This looks like he got a lot closer: but of course the cat also moved. John loves his cat and is going to keep running. By the time he reaches the point where the cat is now, 1.5 miles from the initial starting point, the cat will have moved another 0.25 miles. The new situation looks like this.

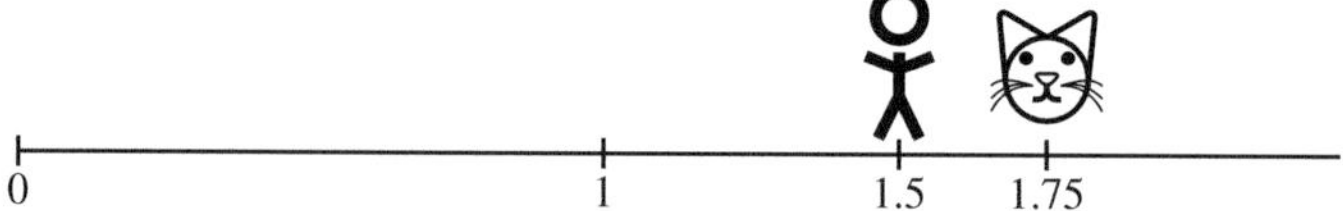

At this point, the paradox is becoming clear: whenever John has reached the point where the cat used to be, the cat has already moved a little bit farther. This will always repeat itself, John will never reach the cat. This is weird because in practice this never seems to cause any actual problems. Maybe this proves that real life is just an illusion? Maybe we only imagine ourselves moving but this is all just a dream since, as we have just seen, all movement is impossible? Zeno, who initially raised this paradox, is again living up to our standard of Greek philosophers having interesting lives: he had a *very unusual* death.

> he plotted to overthrow Nearchus the tyrant but was arrested [...] he was cross-examined [...] Then, saying that he had something to tell him about certain people in his private ear, he laid hold of it with his teeth and did not let go until stabbed to death. (Laertius, [160])

Going back to John and the cat, there is a nice way to resolve the conundrum. We first note that John, running at speed 1, is after t units of time going to be at position $J(t) = t$. The cat, having a head start and running at half the speed, is going to be at $C(t) = t/2 + 1$. Solving the equation

$$J(t) = t = t/2 + 1 = C(t)$$

we quickly see that John will catch the cat after $t = 2$ units of time and exactly 2 miles from the initial starting point. So what is happening in Zeno's paradox? The trick lies in observing that even though it appears that there are infinitely many steps (every time John reaches the point where the cat used to be, the cat has moved on already), these steps get smaller and smaller and smaller. First, John runs distance 1. Then, to catch up with the cat, he has to run distance $1/2$. Then, while running distance $1/2$, the cat – moving at half the speed – manages to travel another distance of $1/4$. Writing down these equations, we see that

$$\text{total distance traveled} = 1 + \frac{1}{2} + \frac{1}{4} + \frac{1}{8} + \cdots = 2.$$

This is quite interesting because it shows that *it is possible for infinitely many numbers to add to a finite number*. Of course, this probably requires some restriction on what we add up: the numbers that we add up have to become eventually quite small. If they

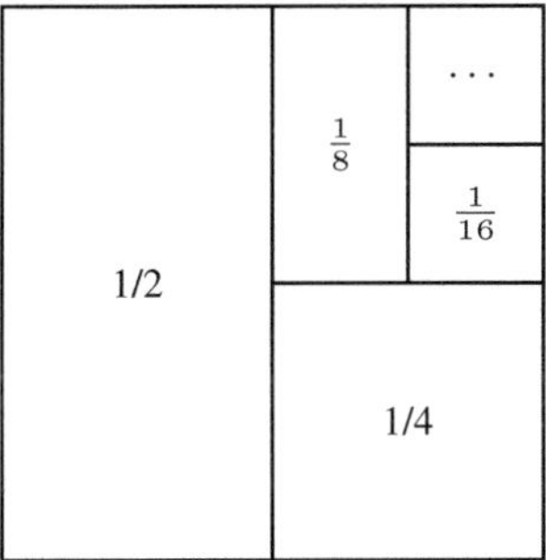

Fig. 8.1: Another explanation why $1/2 + 1/4 + 1/8 + \cdots = 1$.

do not get smaller and there are infinitely many of them, the final result will probably not be a finite number. Having realized this, we see the phenomenon everywhere. For example, the Champernowne constant $0.1234567\ldots$ is clearly a finite number: it is between 0 and 1. However, the way we write this number with digits after the decimal point is really an abbreviation for

$$0.1234567\cdots = \frac{1}{10} + \frac{2}{100} + \frac{3}{1000} + \frac{4}{10000} + \frac{5}{100000} + \cdots$$

which is clearly the sum of infinitely many things summing up to a finite number. This is an interesting phenomenon, and we will have all sorts of fun with it.

8.3 The geometric series

Geometric series were the first infinite sums for which a finite value was determined: pick some number $-1 < q < 1$ and consider the sum

$$S = 1 + q + q^2 + q^3 + q^4 + \cdots$$

We have already seen above that when $q = 1/2$, this sum evaluates to 2.

Theorem (Geometric series). *Let $-1 < q < 1$. Then*

$$1 + q + q^2 + q^3 + \cdots = \frac{1}{1-q}.$$

Proof. Let us say that $1 + q + q^2 + q^3 + \cdots = S$. We do not quite know what S is and hope to find out. Multiplying both sides of the equation by the number q

$$q + q^2 + q^3 + q^4 + \cdots = q \cdot S.$$

However, the quantity on the left-hand side is simply $S - 1$ since

$$S - 1 = q + q^2 + q^3 + \cdots .$$

Therefore $S - 1 = q \cdot S$, which implies the result. $\qquad\square$

The geometric series is presumably the nicest example of an infinite series: it has a particularly nice form and a nice formula. Typically, adding infinitely many things is usually kind of complicated. Often there is no hope of getting a nice formula. The geometric series is so fantastically nice that many proofs of special cases have been given. When writing *Quadrature of the Parabola* in the third century BC, Archimedes has to determine that

$$\frac{1}{4} + \frac{1}{16} + \frac{1}{64} + \frac{1}{256} + \cdots = \frac{1}{3}.$$

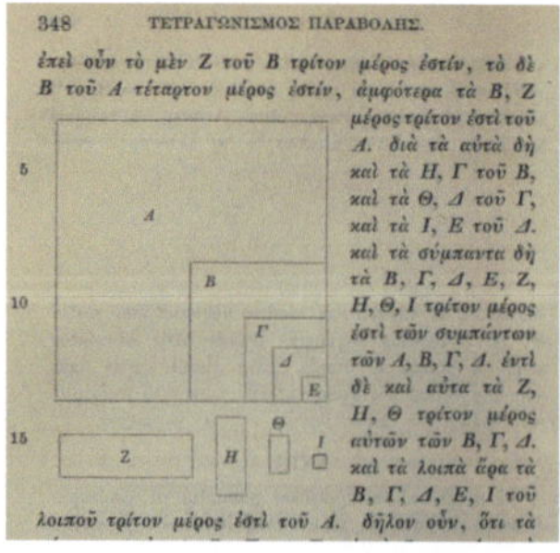

Fig. 8.2: Archimedes summing the geometric series [8].

His method is ingenious, see Fig. 8.2. Dividing the unit square (of area 1) as shown, the sum in question corresponds to shaded areas and for each shaded square we can find two more unshaded squares of the same size: thus the total shaded area is $1/3$ of the total area. Archimedes is quite the legend! John Tzetzes, a Byzantine poet who lived in the 12th century, recalls in his *Books of Histories* [285] the Roman attack on Archimedes' hometown Syracuse in 213 BC. He describes Archimedes setting fire to Roman ships by focusing the rays of the sun using a mirror.

Fig. 8.3: Archimedes' heat ray setting fire to a ship (Giulio Parigi, 1571–1673).

Wise Archimedes, that machinist,
Was a Syracusan by race, an old geometrician,
And driving past seventy-five seasons,
A man who built many mechanical capacities [...]
The old man constructed some sort of six-angled mirror [...]
He set that six-angled mirror in the middle of the rays of the sun
When it was mid-day both in the summer and in the most wintry season.
When the rays, later, were reflected into this,
A fearful fiery kindling was lifted to the vessels,
And reduced them to ashes [...] (John Tzetzes, *Chiliades* [285])

Archimedes did not survive the Roman attack. In a story that has become famous, he was working on some idea and asked an approaching Roman soldier to not disturb his work. The full story is even better and shows Archimedes *fighting back*.

He was bent forward, drawing some mechanical diagram,
But some Roman standing near him dragged him in an attempt to take him prisoner.
And Archimedes, being then wholly engulfed in his diagram,
Not knowing who was dragging him down, said to that man:
"Stand away, o man, from my diagram."
As the Roman was dragging him, Archimedes twisted up, recognizing that he was Roman,
Shouted, "Someone, give me one of my mechanisms."
But the Roman, excited, immediately kills him,
A decayed old man, divine in his works. (John Tzetzes, *Chiliades* [285])

Claudius Marcellus, commander of the Roman forces, *immediately upon learning this, lamented, And illustriously buried this man.* Moreover, *the killer of Archimedes, I think, Marcellus kills with an axe* [285].

8.4 The harmonic series

For a long time, the geometric series was the only known example of an infinite sum. This changed in the 14th century when new examples started to emerge.

Theorem (The harmonic series diverges). *We have*

$$\frac{1}{1} + \frac{1}{2} + \frac{1}{3} + \frac{1}{4} + \frac{1}{5} + \frac{1}{6} + \cdots = \infty.$$

We could start by looking at the first few terms: even though the numbers keep growing, they grow *very* slowly. If we sum up a number of terms roughly comparable to the number of atoms in the universe, we do not even exceed 200

$$\frac{1}{1} + \frac{1}{2} + \frac{1}{3} + \frac{1}{4} + \cdots + \frac{1}{10^{80}} = 184.784\ldots$$

The idea behind the proof is pretty simple and goes back to Nicolas Oresme (1320–1382), one of the most original thinkers of the 14th century – and remarkably modern! In his *Book of the Heavens and the Earth*, he ponders the question of whether the Earth moves and the sky is fixed or whether the Earth is fixed and the sky rotates around the earth and arrives at thoughts about the relativity of movement that could be straight out of Einstein

> Now, let us imagine and assume to be possible that God [...] created two bodies [...] which we will designate a and b. Let us suppose a moves and b rests; then a and b would change their relative position completely just as though b moved and a rested. (Oresme, *Le Livre du ciel et du monde*, (93 c-d) [217])

Proof. Oresme's argument is short and simple. How large is

$$\frac{1}{17} + \frac{1}{18} + \cdots + \frac{1}{32} \qquad ?$$

We could compute it exactly but there is an easy argument to produce an under-estimate: there are 16 terms and each of them is at least $1/32$. Therefore

$$\frac{1}{17} + \frac{1}{18} + \cdots + \frac{1}{32} \geq 16 \cdot \frac{1}{32}.$$

We can use the same argument for other powers of 2

$$\frac{1}{33} + \frac{1}{34} + \cdots + \frac{1}{64} \geq 32 \cdot \frac{1}{64} = \frac{1}{2}$$
$$\frac{1}{65} + \frac{1}{66} + \cdots + \frac{1}{128} \geq 64 \cdot \frac{1}{128} = \frac{1}{2}$$

and since there are infinitely many powers of 2, we add something that is at least as large as $1/2$ an infinite number of times. That is surely infinite. $\qquad\square$

Proof by Cohen–Knight. Many other arguments exist: here is a very funny one due to Teresa Cohen and William Knight [59]. Suppose

$$1 + \frac{1}{2} + \frac{1}{3} + \cdots = S \qquad \text{is finite.}$$

Then we can divide it into sums with even terms and sums with odd terms

$$S = \left(1 + \frac{1}{3} + \frac{1}{5} + \dots\right) + \left(\frac{1}{2} + \frac{1}{4} + \frac{1}{6} + \cdots\right)$$

and we see that the first sum, containing the odd terms, is larger than the second term because $1 > 1/2$ and $1/3 > 1/4$ and so on. However, dividing the entire harmonic series by 2, we see

$$\frac{1}{2} + \frac{1}{4} + \frac{1}{6} + \cdots = \frac{1}{2}\left(1 + \frac{1}{2} + \frac{1}{3} + \cdots\right) = \frac{S}{2}$$

implying that both expressions have exactly the same size. This is a contradiction. $\square$

8.5 Combining geometric and harmonic series

There exists an absolutely miraculous combination of the convergence of the geometric series and the divergence of the harmonic series. The prime numbers

$$2, 3, 5, 7, 11, 13, 17, 19, 23, 29, 31, \ldots$$

are the numbers that are only divisible by themselves and 1. Phrased differently, a number $n \in \mathbb{N}$ with $n \geq 2$ is prime if and only if $a, b \in \mathbb{N}$ and $n = a \cdot b$ then either $a = 1$ or $b = 1$. A fundamental fact, going back to Euclid, is that there are infinitely many prime numbers.

Theorem. *There exist infinitely many prime numbers.*

Proof. There are many proofs of this fact but few are so astonishing as this proof that was given by Euler [80] in 1744. It only uses the fact that each integer is a product of prime numbers and the geometric series

$$1 + \frac{1}{2} + \frac{1}{4} + \frac{1}{8} + \frac{1}{16} + \cdots = 2 \tag{8.1}$$

$$1 + \frac{1}{3} + \frac{1}{9} + \frac{1}{27} + \frac{1}{81} + \cdots = \frac{4}{3} \tag{8.2}$$

and, more generally, for any prime number p

$$1 + \frac{1}{p} + \frac{1}{p^2} + \frac{1}{p^3} + \cdots = \frac{p}{p-1}.$$

This follows from the formula for the geometric series with $q = 1/p$. Looking now at $1 + 1/2 + 1/4 + 1/8 + \cdots$, we recognize a part of the harmonic series: it's the harmonic series restricted to numbers that are powers of 2. Euler's fantastic idea is now to multiply the two equations (8.1) and (8.2).

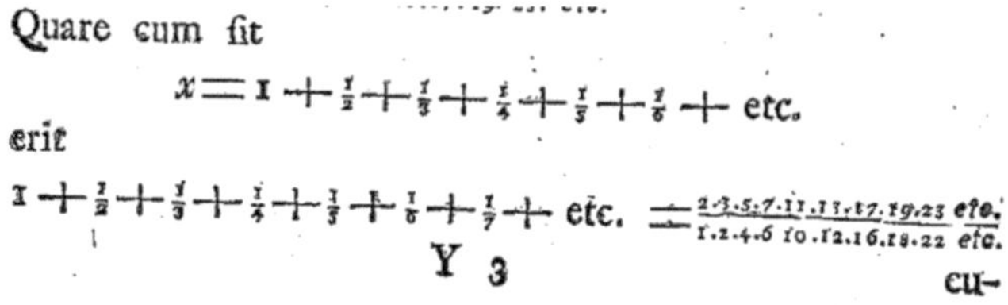

Quare cum fit

$$x = 1 + \tfrac{1}{2} + \tfrac{1}{3} + \tfrac{1}{4} + \tfrac{1}{5} + \tfrac{1}{6} + \text{ etc.}$$

erit

$$1 + \tfrac{1}{2} + \tfrac{1}{3} + \tfrac{1}{4} + \tfrac{1}{5} + \tfrac{1}{6} + \tfrac{1}{7} + \text{ etc.} = \frac{2 \cdot 3 \cdot 5 \cdot 7 \cdot 11 \cdot 13 \cdot 17 \cdot 19 \cdot 23 \text{ etc.}}{1 \cdot 2 \cdot 4 \cdot 6 \cdot 10 \cdot 12 \cdot 16 \cdot 18 \cdot 22 \text{ etc.}}$$

Fig. 8.4: *If we set ... then*: Euler's 1744 proof [80] that there are infinitely many primes by combining geometric series and harmonic series.

The right-hand side is easy to multiply, we get $2 \cdot 4/3 = 8/3$. Multiplying the left-hand sides looks more complicated but if we just start and multiply term by term as we are used to, we arrive at

$$\frac{8}{3} = 1 + \frac{1}{2} + \frac{1}{3} + \frac{1}{4} + \frac{1}{6} + \frac{1}{8} + \frac{1}{9} + \frac{1}{12} + \frac{1}{16} + \frac{1}{18} + \cdots$$

This may look complicated at first until we realize: this is the harmonic series restricted to numbers that are only divisible by 2 and 3. Euler's argument continues. Multiplying this by the geometric series for $p = 5$

$$\frac{2}{2-1} \cdot \frac{3}{3-1} \cdot \frac{5}{5-1} = 1 + \frac{1}{2} + \frac{1}{3} + \frac{1}{4} + \frac{1}{5} + \frac{1}{6} + \frac{1}{8} + \frac{1}{9} + \frac{1}{10} + \frac{1}{12} + \cdots$$

and we get the harmonic series restricted to numbers that are divisible by 2, 3 or 5. At this point, Euler argues by contradiction: suppose there are only finitely many prime numbers. Let us call them $p_1, p_2, \ldots, p_k$. Maybe k is very large, much larger than the number of atoms in the Universe. However, no matter how large k is, if it is finite, then we can multiply all k geometric series together and call the result x

$$x = \left(1 + \frac{1}{2} + \frac{1}{4} + \ldots\right)\left(1 + \frac{1}{3} + \frac{1}{9} + \ldots\right) \cdots \left(1 + \frac{1}{p_k} + \frac{1}{p_k^2} + \ldots\right).$$

On the one hand, using the formula for the geometric series, we have that

$$x = \frac{2}{2-1}\frac{3}{3-1} \cdots \frac{p_k}{p_k - 1}.$$

We do not know what that number is, it depends a bit on k and what the prime numbers are. However, x is certainly a finite number: we have $p/(p-1) \leq 2$ for all prime numbers and thus $x \leq 2^k$. On the other hand, multiplying all the terms, we get the harmonic series without any restrictions

$$x = 1 + \frac{1}{2} + \frac{1}{3} + \frac{1}{4} + \frac{1}{5} + \cdots = \infty.$$

That is a contradiction, there have to be infinitely many prime numbers. □

8.6 The Swineshead series

The first examples of infinite series (other than the geometric series) are the harmonic series and a series due to Richard Swineshead. Richard Swineshead (also Richard Suiseth or Richard Suisset and John Suisset) was an English mathematician and part of the *Oxford calculators*.

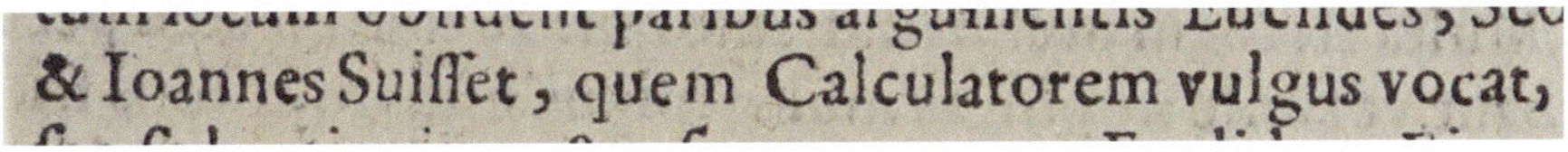

& Ioannes Suiſſet, quem Calculatorem vulgus vocat,

Fig. 8.5: Cardano in Book 16 of *De Subtilitate* [47] listing *John Suisset, commonly known as 'The Calculator'* next to Archimedes and Euclid.

Swineshead is almost completely forgotten. The Italian polymath Gerolamo Cardano (1501–1576) lists Swineshead/Suisset in his list of the 12 greatest scholars of all time; the others being, following Cardano's ordering, Archimedes (who is *not only first but inimitable*), Aristotle, Euclid, Duns Scotus, John Suisset, the geometer Apollonius of Perga, the Pythagorean Archytas of Tarentum, who founded modern music theory, Al-Khwarizmi, whose name is immortalized in 'Algebra', Al-Kindi, Jabir ibn Hayyan (*Geber of Spain*), Galen of Pergamon and the Roman architect Vitruvius. Swineshead is also mentioned in Robert Burton's 1621 *The Anatomy of Melancholy* as

> Suisset the calculator, *qui pene modum excessit humani ingenii* [who almost exceeded the limits of human talent] (Robert Burton, *The Anatomy of Melancholy* [42])

Swineshead is not well known these days, neither is Robert Burton's masterpiece *The Anatomy of Melancholy, What it is: With all the Kinds, Causes, Symptomes, Prognostickes, and Several Cures of it. In Three Maine Partitions with their several Sections, Members, and Subsections. Philosophically, Medicinally, Historically, Opened and Cut Up* [42]. It aims to be the most comprehensive treatise on depression ever compiled and may well have succeeded in this task. It is a remarkable work with a long list of admirers. Samuel Johnson said that this *was the only book that ever took him out of bed two hours sooner than he wished to rise* [33]. Samuel Beckett noted fragments from Burton in his notebook, including *plures crapula quam gladius* (gluttony kills more than the sword) as well as a warning that parsnips *trouble the mind, sending gross fumes to the brain* [155]. Burton's book has an opinion about everything. This includes the topic of beans where he echoes Pythagoras *beans, peas, vetches, &c., they fill the brain (saith Isaac) with gross fumes, breed black thick blood, and cause troublesome dreams. And therefore, that which Pythagoras said to his scholars of old, may be for ever applied to melancholy men, A fabis abstinete, eat no peas, nor beans*. Bread can be dangerous (*Bread that is made of baser grain, as peas, beans, oats, rye, or over-hard baked, crusty, and black, is often spoken against, as causing melancholy juice and wind.*), spices are to be avoided (*Spices cause hot and head melancholy, and are for that cause forbidden by our physicians to such men as are inclined to this malady, as pepper, ginger, cinnamon, cloves, mace, dates, &c. honey and sugar*). Beware of herbs (*Amongst herbs to be eaten I find gourds, cucumbers, coleworts, melons, disallowed, but especially cabbage. It causeth troublesome dreams, and sends up black vapours to the brain*). The list goes on. However, Burton is exceedingly fond of mathematics: *What more pleasing studies can there be than the mathematics, theoretical or practical parts? [...] such is the excellency of these studies, that all those ornaments and childish bubbles of wealth, are not worthy to be compared to them* [42]. As for melancholy in general, his main advice is simple

> Only take this for a corollary and conclusion, as thou tenderest thine own welfare in this, and all other melancholy, thy good health of body and mind, observe this short precept, give not way to solitariness and idleness. "Be not solitary, be not idle." (Robert Burton, *The Anatomy of Melancholy* [42])

Returning to Richard Suiseth, the *Oxford calculator*, he managed to compute another series that is neither geometric nor harmonic.

Theorem (Richard Swineshead). *We have*

$$\frac{1}{2} + \frac{2}{4} + \frac{3}{8} + \frac{4}{16} + \frac{5}{32} + \frac{6}{64} + \cdots = 2.$$

Proof. Swineshead has a complicated proof using only words that can be found in Boyer [35]. We simplify things a little and start by calling the correct answer $S = 1/2 + 2/4 + 3/8 + \cdots$. The series can be written in a funny way

$$
\begin{aligned}
S = {} & \frac{1}{2} + \frac{1}{4} + \frac{1}{8} + \frac{1}{16} + \frac{1}{32} + \frac{1}{64} + \cdots \\
& + \frac{1}{4} + \frac{1}{8} + \frac{1}{16} + \frac{1}{32} + \frac{1}{64} + \cdots \\
& + \frac{1}{8} + \frac{1}{16} + \frac{1}{32} + \frac{1}{64} + \cdots
\end{aligned}
$$

and so on.

The first row sums to 1, that is just the geometric series. The second row can be written as half the first row, that gives $1/2$. The third row is half the second row, that is $1/4$, and we recover the geometric series $1 + \frac{1}{2} + \frac{1}{4} + \frac{1}{8} + \cdots = 2$. $\qquad \square$

The Swineshead series is really just two geometric series in disguise. Going through the argument, we also arrive at the following result (sometimes attributed to Oresme).

Theorem. *If* $-1 < q < 1$, *then*

$$q + 2q^2 + 3q^3 + 4q^4 + 5q^5 + \cdots = \frac{q}{(1-q)^2}.$$

Proof. We do not know the answer, so let us denote it by X. Using the geometric series gives us $q+q^2+q^3=q/(1-q)$. Subtracting this from the full series, we get

$$\begin{aligned} X &= \frac{q}{1-q} + q^2 + 2q^3 + 3q^4 + 4q^5 + \cdots \\ &= \frac{q}{1-q} + q\left(q + 2q^2 + 3q^3 + 4q^4 + \cdots\right) = \frac{q}{1-q} + qX. \end{aligned}$$

Solving for X gives us the desired answer. □

Several such results are possible. Here is another one.

Theorem. *If* $-1<q<1$, *then*

$$1 + 2q + 3q^2 + 4q^3 + 5q^4 + 6q^5 + \cdots = \frac{1}{(1-q)^2}.$$

Proof. One can prove it directly using the previous result. However, one could also multiply the geometric series by itself and obtain

$$\left(1 + q + q^2 + q^3 + \ldots\right)\left(1 + q + q^2 + q^3 + \ldots\right) = \frac{1}{(1-q)^2}.$$

Now multiplying out both sides, we have to see how many ways we can arrive at q^n: we can multiply 1 by q^n, we can multiply q by q^{n-1}, we can multiply q^2 by q^{n-2} and so on until we multiply q^n by 1, for a total of $(n+1)$ possibilities. Thus the term q^n on the left-hand side has coefficient $n+1$, as desired. □

8.7 The Leibniz series

One of the people who appreciated Swineshead was the German polymath Gottfried Wilhelm von Leibniz (1646–1716). Referring to Swineshead as Suisse he writes

> There was once a Suisse, who did mathematics belonging to scholasticism; his works are little known, but what I have seen of them seemed to me profound and relevant. (Letter from Leibniz to Remond de Montmorency, August 26, 1714 [167])

Leibniz worked as mathematician, philosopher, scientist and diplomat. In mathematics, he is most famous as the co-inventor of Calculus together with Isaac Newton.

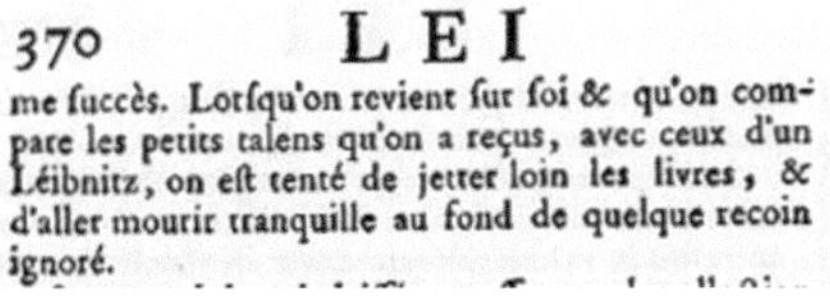
370 LEI

me ſuccès. Lorſqu'on revient ſur ſoi & qu'on compare les petits talens qu'on a reçus, avec ceux d'un Léibnitz, on eſt tenté de jetter loin les livres, & d'aller mourir tranquille au fond de quelque recoin ignoré.

Fig. 8.6: *Encyclopédie* entry for Leibniz: *If one looks back at oneself and compares the little talents that one has received with those of a Leibnitz, one is tempted to throw away the books and to go die quietly in some forgotten corner.*

To get an impression of the impact that Leibniz had, we visit the corresponding entry in the *Encyclopédie*, the massive encyclopedia published between 1751 and 1772. The entry *Léibnitzianisme* was (presumably) written by the French philosopher Denis Diderot The Encyclopédie makes for good reading! Regarding Richard Swineshead (under '*scholastiques*') it says *they hastened to make him suspect of heterodoxy. How could a man know algebra, and fill his physics with unintelligible characters, without being a magician or an atheist? [...] If our hypocrites, our false devotees dared it, they would condemn to the fire anyone who understands the mathematical principles of Newton's philosophy, or who possesses a fossil.* Needless to say, the Encyclopédie was not universally liked (see Fig. 8.7).

Encyclopedie, ou Dictionaire raisonné des Sciences, des Artes, & des Metiers. Par une Societé des Gens des Lettres &c. *Brevi Clem. PP. XIII.* 3. *Septemb.* 1759.

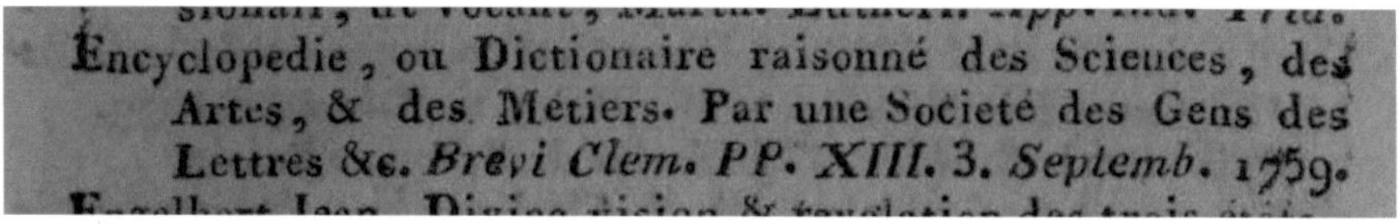

Fig. 8.7: The *Encyclopédie* as listed in the 1819 *Index of Forbidden Books* (Index librorum prohibitorum), a list of books that Catholics were not allowed to read.

Theorem (Leibniz). *We have*

$$\frac{1}{1\cdot 2}+\frac{1}{2\cdot 3}+\frac{1}{3\cdot 4}+\frac{1}{4\cdot 5}+\cdots = 1.$$

Proof. This series has another nice trick which had also been observed by Evangelista Torricelli and Pietro Mengoli. Mengoli studied

$$\frac{2}{2\cdot 3}+\frac{2}{3\cdot 4}+\frac{2}{4\cdot 5}+\cdots = 1,$$

which is seen to be equivalent to the Leibniz series. Note that

$$\frac{1}{1}-\frac{1}{2}=\frac{1}{1\cdot 2} \quad\text{and}\quad \frac{1}{2}-\frac{1}{3}=\frac{1}{2\cdot 3} \quad\text{and}\quad \frac{1}{3}-\frac{1}{4}=\frac{1}{3\cdot 4}$$

and so on. Summing everything on the left-hand side, we get

$$\frac{1}{1}-\frac{1}{2}+\frac{1}{2}-\frac{1}{3}+\frac{1}{3}-\frac{1}{4}+\frac{1}{4}-\frac{1}{5}+\cdots = 1$$

because the $-1/2$ from the first summand cancels with the $+1/2$ from the second summand, $-1/3$ from the second summand cancels with the $+1/3$ from the third and so on. Summing the right-hand side is the series we are interested in. □

This trick is so good, we can apply it to all sorts of other settings. Considering the Fibonacci numbers starting at 1, meaning $1, 2, 3, 5, 8, 13, 21, 34, \ldots$, one has

$$\frac{1}{1\cdot 2}+\frac{1}{2\cdot 3}+\frac{2}{3\cdot 5}+\frac{3}{5\cdot 8}+\frac{5}{8\cdot 13}+\frac{8}{13\cdot 21}+\cdots = 1$$

since we can also write it as

$$\frac{2-1}{1\cdot 2} + \frac{3-2}{2\cdot 3} + \frac{5-3}{3\cdot 5} + \frac{8-5}{5\cdot 8} + \frac{13-8}{8\cdot 13} + \cdots = 1.$$

Theorem. *If $0 < a_1, a_2, \ldots,$ is a sequence of increasing real numbers, then*

$$\frac{a_2 - a_1}{a_1 \cdot a_2} + \frac{a_3 - a_2}{a_2 \cdot a_3} + \frac{a_4 - a_3}{a_3 \cdot a_4} + \cdots = \frac{1}{a_1}.$$

Proof. The proof works just as above: note that

$$\frac{1}{a_1} - \frac{1}{a_2} = \frac{a_2 - a_1}{a_1 \cdot a_2} \quad \text{and} \quad \frac{1}{a_2} - \frac{1}{a_3} = \frac{1}{a_2 \cdot a_3}$$

and so on. Summing over the left-hand side, we get

$$\left(\frac{1}{a_1} - \frac{1}{a_2}\right) + \left(\frac{1}{a_2} - \frac{1}{a_3}\right) + \cdots = \frac{1}{a_1}$$

and summing over the right-hand side gives us the desired infinite series. $\square$

Leibniz also studied other infinite series. One example is a series that was first discovered by the Indian mathematician Madhava of Sangamagrama (1350–1425), later rediscovered by James Gregory in 1671 and by Leibniz in 1673. It is

$$1 - \frac{1}{3} + \frac{1}{5} - \frac{1}{7} + \frac{1}{9} - \frac{1}{11} + \frac{1}{13} - \cdots = \frac{\pi}{4}.$$

This is a bit of a magical statement: what do the odd numbers have to do with π, the area of unit circle? There is a nice and simple proof that uses only elementary calculus. Since the argument is very short, we include it for the sake of people who really like integrals: the secret lies in the derivative of the arctangent function is $\arctan'(x) = 1/(1 + x^2)$ and the basic fact that $\pi/4 = \arctan(1)$. Therefore

$$\frac{\pi}{4} = \arctan(1) = \arctan(1) - \arctan(0) = \int_0^1 \frac{1}{1+x^2}\,dx.$$

Using the geometric series

$$\frac{1}{1+x^2} = 1 - x^2 + x^4 - x^6 + x^9 - \cdots$$

and

$$\int_0^1 1 - x^2 + x^4 - x^6 + x^9 - \cdots\,dx = \int_0^1 1 - \int_0^1 x^2\,dx + \int_0^1 x^4\,dx - \cdots$$
$$= 1 - \frac{1}{3} + \frac{1}{5} - \frac{1}{7} + \cdots.$$

8.8 The Basel problem

This Basel problem was raised by the Italian mathematician Pietro Mengoli, who had many nice ideas. Two of them already appeared above and we will revisit them through his eyes. The first is the divergence of the harmonic series (see [23]).

Theorem. *We have*

$$\frac{1}{1} + \frac{1}{2} + \frac{1}{3} + \frac{1}{4} + \frac{1}{5} + \frac{1}{6} + \cdots = \infty.$$

Mengoli's Proof. The inequality is based on the fact that the sum of three consecutive terms is larger than 3 times the middle term, meaning

$$\frac{1}{n-1} + \frac{1}{n} + \frac{1}{n+1} > \frac{3}{n}$$

which can be seen by rearranging the terms. Using this inequality, we can now do a proof by contradiction. Suppose

$$\frac{1}{1} + \frac{1}{2} + \frac{1}{3} + \frac{1}{4} + \frac{1}{5} + \frac{1}{6} + \cdots = S$$

is finite. We can then rewrite the sum and use the inequality to deduce

$$\begin{aligned}
S &= \frac{1}{1} + \left(\frac{1}{2} + \frac{1}{3} + \frac{1}{4}\right) + \left(\frac{1}{5} + \frac{1}{6} + \frac{1}{7}\right) + \cdots \\
&> \frac{1}{1} + \frac{3}{3} + \frac{3}{6} + \frac{3}{9} + \cdots \\
&= \frac{1}{1} + \frac{1}{1} + \frac{1}{2} + \frac{1}{3} + \cdots = S + 1.
\end{aligned}$$

However, no number S satisfies $S > S + 1$ and that is a contradiction. $\square$

Mengoli in his 1650 book *Novae quadraturae arithmeticae*, after discussing the Leibniz series, raises another question.

> Having successfully dealt with the contemplation of this arrangement of fractions, I proceed to another arrangement, in which the individual units are denominated by square numbers. (Mengoli [194])

The fruits of this work are considered to be the Theorems, those first of all which are demonstrated in the first book, and in addition the Propositions that follow [194]. He then reports one more Leibniz-type series: for example, he says, if you take over the inverse squares plus twice the side, then this sums to $3/4$, by which he means

$$\frac{1}{1^2 + 2 \cdot 1} + \frac{1}{2^2 + 2 \cdot 2} + \frac{1}{3^2 + 2 \cdot 3} + \frac{1}{4^2 + 2 \cdot 4} + \cdots + \frac{1}{n^2 + 2n} + \cdots = \frac{3}{4}.$$

He manages to solve the problem with quantities that behave approximately like the square, but his original problem

$$\frac{1}{1} + \frac{1}{4} + \frac{1}{9} + \frac{1}{16} + \frac{1}{25} + \frac{1}{36} + \cdots = \ ?$$

remains unsolved. Mengoli says that it is worth investigating, but that a more industrious mind is required. The number, if finite, is clearly larger than 1. There is a simple comparison trick that allows us to show that the number is less than 2 (and thus, in particular, finite). The trick is based on the idea of reducing the series to the Leibniz series that we already analyzed above by arguing as follows

$$\begin{aligned}
\frac{1}{1} + \frac{1}{4} + \frac{1}{9} + \frac{1}{16} + \cdots &= \frac{1}{1} + \frac{1}{2 \cdot 2} + \frac{1}{3 \cdot 3} + \frac{1}{4 \cdot 4} + \cdots \\
&\leq \frac{1}{1} + \frac{1}{1 \cdot 2} + \frac{1}{2 \cdot 3} + \frac{1}{3 \cdot 4} + \cdots = 1 + 1 = 2.
\end{aligned}$$

The problem was finally resolved by Leonhard Euler in 1735, who showed

$$\frac{1}{1^2} + \frac{1}{2^2} + \frac{1}{3^2} + \frac{1}{4^2} + \cdots = \frac{\pi^2}{6}.$$

Euler's proof, which we will not discuss here, works in a slightly more general setting and can also be used to prove $1/1^4 + 1/2^4 + 1/3^4 + \cdots = \pi^4/90$. This suggests a nice pattern that generalizes: we take k-th powers of the integers on the left-hand side and get a π^k on the right-hand side. However, this pattern (probably) fails for $k = 3$. We still do not have a formula for

$$\frac{1}{1} + \frac{1}{2^3} + \frac{1}{3^3} + \frac{1}{4^3} + \cdots = 1.20205690315959\ldots$$

We do not know what that number is! For a long time it was not even clear whether it might be a rational number. In 1979, Apéry [6] showed that the number is irrational. His proof seemed to rely on very strange facts, it was difficult to believe that it could be true. When Apéry was asked where he had gotten one of the strange equations he had used, he supposedly replied *They grow in my garden*, which did not really convince people [174]. There is a fascinating issue here: maybe this number, $\zeta(3)$, simply cannot be described in terms of elementary quantities? There are many ways of writing it, for example

$$\frac{1}{1} + \frac{1}{2^3} + \frac{1}{3^3} + \cdots = \int_0^1 \int_0^1 \int_0^1 \frac{dx\,dy\,dz}{1 - xyz} = \frac{5}{2} \sum_{k=1}^{\infty} (-1)^{k-1} \frac{k!^2}{(2k)! k^3} = \cdots$$

but maybe having a nice formula like $\sqrt{3 + \sin(1)}$ is asking too much? Using the standard abbreviation in terms of the Riemann ζ-function

$$\zeta(k) = 1 + \frac{1}{2^k} + \frac{1}{3^k} + \frac{1}{4^k} + \frac{1}{5^k} + \cdots$$

we know that $\zeta(1) = \infty$ (that's the harmonic series), we know that $\zeta(2) = \pi^2/6$ (Euler's solution of the Basel problem), and we know almost nothing about $\zeta(3)$.

Euler provided explicit formulas for $\zeta(k)$ when k is an even integer, and we know very little about $\zeta(3), \zeta(5), \ldots$. However, there is a fun fact that we do know.

Theorem. *We have*

$$(\zeta(2) - 1) + (\zeta(3) - 1) + (\zeta(4) - 1) + \cdots = 1.$$

Proof. The trick is to sum everything in a different order: then the geometric series is useful. Summing only over the powers of 2, we get

$$\frac{1}{2^2} + \frac{1}{2^3} + \frac{1}{2^4} + \cdots = \frac{1}{2}.$$

Summing, more generally, over the powers of m, we have

$$\frac{1}{m^2} + \frac{1}{m^3} + \frac{1}{m^4} + \cdots = \frac{1}{m^2}\left(1 + \frac{1}{m} + \ldots\right) = \frac{1}{m^2}\frac{m}{m-1} = \frac{1}{(m-1)m}.$$

Then, however, summing this over $m = 2, 3, 4, \ldots$ is simply Leibniz' series that we already evaluated before and the proof is complete. $\qquad\square$

8.9 An impossible shape: Torricelli's trumpet

These two results, the unboundedness of the harmonic series and the boundedness of $1/1^2 + 1/2^2 + 1/3^2 + \cdots$, allow us to prove a surprising result originally due to Torricelli (1608–1647), who also invented the barometer. In connection with that discovery, he is famous for a poetic description that is worth remembering

We live submerged at the bottom of an ocean of air. (Letter, Torricelli to Ricci, 1644 [280])

The letter continues in a more scientific manner and specifies that the air near the earth's surface *weighs about a four-hundredth part of the weight of water*. He also showed in *De dimensione parabolae* that if we are given a sequence of positive terms $a_1, a_2, \ldots, a_n, \ldots$ that get closer and closer to 0, then

$$(a_1 - a_2) + (a_2 - a_3) + (a_3 - a_4) + \cdots = a_1,$$

which is the trick we already saw above several times. Another discovery of Torricelli is a very unusual geometric shape shown in Fig. 8.8. One of our basic intuitions is that some shape has to be either finite or infinite. If it is finite, then it has finite volume and also finite surface area. If it is infinite, then it has infinite volume and infinite surface area. As it turns out, things are not quite so simple, and Torricelli's trumpet is the first natural example.

Theorem (Evangelista Torricelli). *There exists a shape in three-dimensional space which has finite volume and infinite surface area.*

Proof. The proof is nice because it is completely constructive: we stack an infinite number of cylinders on top of each other. They all have constant height 1, the only thing that changes is the radius. The radius of the first cylinder is 1, the radius of the second cylinder is 1/2, the radius of the third cylinder is 1/3 and so on.

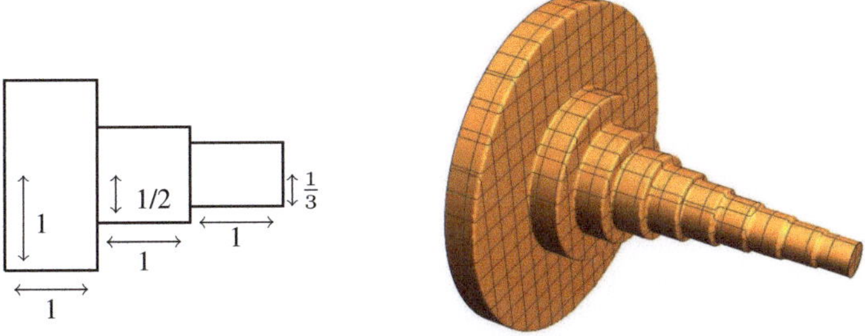

Fig. 8.8: Left: First three layers of Torricelli's trumpet seen from the side, cylinders of constant height 1 with radius $1, 1/2, 1/3$. Right: the shape in three dimensions.

Volume. We have to compute volume and surface area. The volume of a single cylinder with radius r and height h is $r^2\pi h$. The height of our cylinders is always 1, their radii are $1, 1/2, 1/3, \ldots$. Therefore our shape has total volume (recalling the solution of the Basel problem)

$$\pi + \frac{\pi}{4} + \frac{\pi}{9} + \frac{\pi}{16} + \cdots = \pi\left(1 + \frac{1}{4} + \frac{1}{9} + \ldots\right) = \frac{\pi^3}{6} = 5.1677\ldots$$

We note that Torricelli did not know about the solution of the Basel problem, but he knew (just as we saw above) that the infinite sum is less than 2. This is enough to show that the total volume is $\leq 2\pi \sim 6.28\ldots$ and therefore a finite number.

Surface area. Each cylinder has a top, a bottom, and an outer layer. Instead of computing the full surface area, we only compute the surface area of the outer layer: already that will be infinite. A cylinder with radius r and height h has outer surface area $2\pi r h$. In our case, this means that the total surface area is at least

$$\frac{2\pi}{1} + \frac{2\pi}{2} + \frac{2\pi}{3} + \cdots = \infty,$$

where we used the infinitude of the harmonic series. $\qquad\qquad\square$

At first, this seems like a reasonably friendly computation: we now know that it is possible for volume to be finite while surface area is infinite. The clash with intuition comes when one starts thinking of the trumpet as an actual shape, maybe a container. What happens if we were to pour paint into it?

Painting Paradox. Torricelli's trumpet can be filled with a finite amount of paint but its surface cannot be painted with any finite amount of paint.

This paradox has been **aggressively** debated. The first reactions were quite friendly: Torricelli described the discovery to Bonaventura Cavalieri (1596–1647), who replied

> I received your letter while in bed with fever and gout … but in spite of my illness, I enjoyed the savory fruits of your mind, since I found infinitely admirable that infinitely long hyperbolic solid which is equal to a body finite in all the three dimensions. And having spoken about it to some of my philosophy students, they agreed that it seemed truly marvelous and extraordinary that that could be. (Cavalieri to Torricelli, Dec 17, 1641 [186])

Other people were less excited. Pierre Gassendi (1592–1655), after whom a lunar crater is named, writes somewhat incredulously

> And these are the suppositions from which Mathematicians, within the gates of pure and abstract Geometry and almost constituting a kingdom of their own, weave those famous demonstrations, some so extraordinary that they even exceed credibility, like what the famous Cavalieri and Torricelli showed [...] (Gassendi [105])

The philosopher Thomas Hobbes (1588–1679) was strongly against the use of the word 'infinite' and claims

> we are not able to conceive the ends, and bounds of the thing named; having no conception of the thing, but of our own inability. (Hobbes [134])

To this, the mathematician John Wallis (1626–1703) replied:

> A surface, or solid, may be supposed so constituted, as to be Infinitely Long, but Finitely Great, (the Breadth Continually Decreasing in greater proportion than the Length increaseth) [...]. Such is Torricellio's Solidum Hyperbolicum acutum; and others innumerable, discovered by Dr. Wallis, Monsieur Fermat, and others. But to determine this, requires more Geometry and Logic than Mr. Hobs is Master of. (Wallis, cited in: [186])

This seems unusually aggressive and so is Hobbes' reply:

> I do not remember this of Torricellio, and I think Dr. Wallis does him wrong and Monsieur Fermat too. For, to understand this for sense, it is not required that a man should be a geometrician or a logician, but that he should be mad. (Hobbes, cited in: [186])

If this exchange seems a little bit heated, well, there is a reason. This was part of an ongoing debate between Thomas Hobbes and John Wallis that started in 1655. Instead of going into the details (see [146, 237, 257]), we summarize the titles of some of the published works

1. 1656, Hobbes, *Six lessons to the professors of the mathematiques one of geometry the other of astronomy, in the chaires set up by the noble and learned Sir Henry Savile in the University of Oxford.* The title is easier to understand if one is told that Wallis was the Savilian Professor of Geometry at the University of Oxford starting in 1649. The six lessons are remarkable. Lesson III ('On the Faults that Occur in Demonstration') jumps right in.

   > Euclid divides an angle into right, obtuse and acute. I may ask you as pertinently, what angle it is he so divides? Or, when you divide *animal* into *homo* and *brutum*, what animal that is, which you so divide? You see by this, how absurd your question is. [136]

Lesson VI ('Of manners') even justifies Hobbes' rudeness: Wallis started it.

> and remember Vespasian's law, that it is uncivil to give ill language first, but civil and lawful to return it. [136]

2. 1656, Wallis, *Due correction for Mr Hobbes. Or Schoole discipline, for not saying his lessons right.*
3. 1657, Hobbes, *Marks of the Absurd Geometry, Rural Language, Scottish Church-Politicks, and Barbarisms of John Wallis, Professor of Geometry and Doctor of Divinity, by Thomas Hobbes of Malmesbury*
4. 1661, Wallis *Hobbius Heauton-timorumenos, or A Consideration of Mr Hobbes, his Dialogues* [Heauton-timorumenos is Greek for 'Self-Tormentor']
5. 1662, Hobbes *Considerations upon the reputation, loyalty, manners, & religion of Thomas Hobbes of Malmsbury written by himself*
6. 1666, Hobbes, latin title translating to *On the principles of the reasoning of geometers, where uncertainty and falsity are found; that there is no less in their writings than in the writings of physicists and ethics. Against the pride of the professors of geometry.*

The last book has a particularly wonderful sentence in its preface worth quoting

> Of those who with me have written something about these matters, either I alone am mad, or I alone am not mad. No third opinion can be maintained, unless (as perchance it may seem to some) we are all mad. (Hobbes, [135])

Needless to say, Wallis had a clever answer and wrote *If he be mad he is not likely to be convinced by reason; on the other hand, if we be mad, we are in no position to attempt it* [257]. A little while later Hobbes summarizes the ongoing debate in a book whose title translates to *Mathematical Light. Shaken by the collisions of John Wallis, public professor of Geometry in the most famous academy of Oxford, and Thomas Hobbes of Malmsbury* in which he declares himself the winner of the debate.

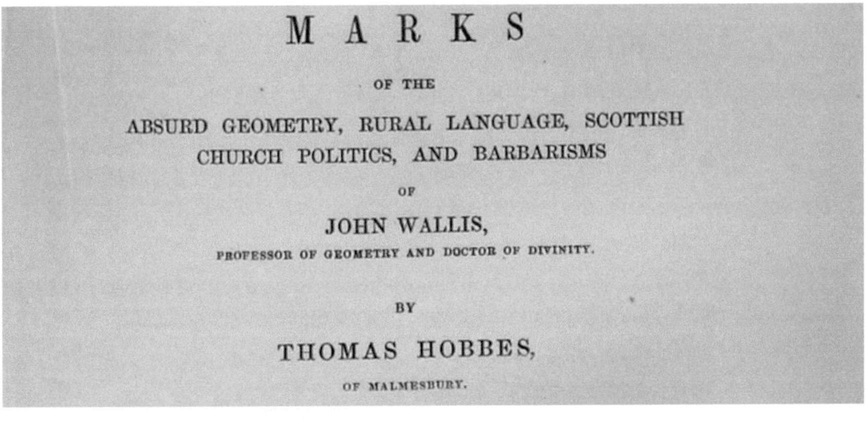

M A R K S

OF THE

ABSURD GEOMETRY, RURAL LANGUAGE, SCOTTISH
CHURCH POLITICS, AND BARBARISMS

OF

JOHN WALLIS,

PROFESSOR OF GEOMETRY AND DOCTOR OF DIVINITY.

BY

THOMAS HOBBES,

OF MALMESBURY.

Fig. 8.9: Possibly the best book title of all time? [136]

Their dispute only ended with Hobbes' death in 1679. The Catholic Church, for a while, forbade all Catholics from reading *any* of Hobbes' works. To have *all* your works listed in the Index of Forbidden Books is a rare honor that he shares with Niccolo Machiavelli, David Hume, Jean-Paul Sartre and Gregorio Leti (1630–1701) (whose 1667 book *Papal Nepotism, or the True Relation of the Reasons Which Impel the Popes to make their Nephews Powerful* may have something to do with it).

The philosophical work of Thomas Hobbes has endured. He is now frequently named, together with John Locke and Jean-Jacques Rousseau, as one of the originators of the social contract theory: broadly speaking that individuals consent to surrender some of their freedoms to an authority (a king or the state) in exchange for protection of their remaining rights through said authority. This makes Hobbes one of the founders of modern political philosophy. If one is surprised to find a rough character like Thomas Hobbes to be the founder of modern political philosophy, it is worth mentioning that, in 1683, John Locke was accused of having been involved with a plan to assassinate King Charles II and had to flee to the Netherlands. The third founder, Jean-Jacques Rousseau, gives a description of his life in his *Confessions*, where he aims to expose his innermost secrets, including him literally exposing himself, though, as he emphasizes, only his behind: *I haunted dark alleys and hidden retreats, where I might be able to expose myself to women in the condition in which I should have liked to have been in their company. What they saw was not an obscene object, I never even thought of such a thing; it was a ridiculous object. The foolish pleasure I took in displaying it before their eyes cannot be described* [247]. Rousseau also wrote a treatise on the nature of education that he never tested: his own children were all sent to the orphanage

> Five children were born of [our] liaison, and all were placed in the Foundling's Hospital, and with so little thought of the possibility of their identification that I did not even keep a record of their dates of birth or gender. (Rousseau to Maréchale de Luxembourg [154])

Rousseau, during one of his endless amorous adventures, once encountered unusual advice, *I attempted to place myself by her side: she withdrew to a sofa, rose from it the next moment, and fanning herself as she walked about the chamber, said to me in a reserved and disdainful tone of voice, "Zanetto, lascia le donne, e studia la matematica." [Leave women and study mathematics.]* [247].

8.10 Exercises

1. Find out where John catches his cat and then convince a friend that John will be unable to reach his cat, and that reality, as Zeno teaches, is an illusion.
2. Determine
$$\frac{1}{7} + \frac{2}{7} + \frac{1}{7^2} + \frac{2}{7^2} + \frac{1}{7^3} + \frac{2}{7^3} + \cdots = \ ?$$

3. Consider the geometric series $1 + q + q^2 + q^3 + \cdots = 1/(1-q)$, which is a true statement for all $-1 < q < 1$. Plug in $q = -0.99$, $q = -0.999$, $q = -0.9999$ and see what happens.

4. Use Oresme's proof to show that

$$1 + \frac{1}{2} + \frac{1}{3} + \frac{1}{4} + \cdots + \frac{1}{n} \geq \frac{\log_2 n}{10}.$$

5. Reexamine Euler's proof of the infinitude of prime numbers. Show that it implies that if $p_1, \ldots, p_k$ are the prime numbers $\leq n$, then

$$1 + \frac{1}{2} + \frac{1}{3} + \frac{1}{4} + \cdots + \frac{1}{n} \leq \frac{p_1}{p_1 - 1} \frac{p_2}{p_2 - 1} \cdots \frac{p_k}{p_k - 1} \leq 2^k.$$

Use the previous exercise to prove that the number k of prime numbers that are $\leq n$ is $k \geq c \log \log n$ for some universal constant $c > 0$.

6. Prove or disprove: for all possible choices of signs

$$1 \pm \frac{1}{4} \pm \frac{1}{9} \pm \frac{1}{16} \pm \cdots \qquad \text{is always finite.}$$

7. Spend some time learning about the life of Leibniz. Bonus point if, unlike Diderot, you can do so use without feeling the temptation to afterwards *go die quietly in the dark of some forgotten corner.*

8. Determine the value of the sum

$$\frac{2}{1 \cdot 3} + \frac{2}{3 \cdot 5} + \frac{2}{5 \cdot 7} + \frac{2}{7 \cdot 9} + \cdots =$$

9. We saw a proof that $1 + 1/4 + 1/9 + 1/16 \leq 2$. Adapt the proof to show that it is actually smaller than 1.9.

10. Wallis writes that *A surface, or solid, may be supposed so constituted, as to be Infinitely Long, but Finitely Great, (the Breadth Continually Decreasing in greater proportion than the Length increaseth) [...]. Such is Torricellio's Solidum Hyperbolicum acutum; and others innumerable.* Try to find another one.

11. Resolve the Painting Paradox. Or has mathematics found a way to paint infinite surfaces with a finite amount of paint?

12. Oresme's argument has been extended, almost 500 years later, to a very useful technique (the 'Cauchy Condensation Test'). If $a_1 \geq a_2 \geq a_3 \geq \cdots \geq 0$, then $a_1 + a_2 + a_3 + a_4 + \cdots$ is finite if and only of $a_1 + 2 \cdot a_2 + 4 \cdot a_4 + 8 \cdot a_8 + 16 \cdot a_{16} + \cdots$ is finite. The argument simply uses that

$$2^n \cdot a_{2^{n+1}} \leq a_{2^n+1} + a_{2^n+2} + \cdots + a_{2^{n+1}} \leq 2^n \cdot a_{2^n}.$$

Fill in the details and use the Cauchy Condensation Test to prove that

$$\frac{1}{1^\alpha} + \frac{1}{2^\alpha} + \frac{1}{3^\alpha} + \frac{1}{4^\alpha} + \cdots \qquad \text{is finite if and only if } \quad \alpha > 1.$$

Chapter 9
Incorrectly Adding Infinitely Many Things

This section is special: it is devoted to mistakes! The lesson to be learned is that it is important *how* one asks questions. A question may be misleading, it may be suggestive, it may implicitly presuppose a statement. Diogenes Laertius gives a beautiful example ('When did you stop beating your father?')

> once when Alexinus asked him whether he had left off beating his father, he said, "I have not beaten him, and I have not left off"; and when he said further that he ought to put an end to the doubt by answering explicitly yes or no, "It would be absurd," he rejoined, "to comply with your conditions, when I can stop you at the entrance." (Diogenes Laertius [160])

It seems hard to believe that this has any implications for mathematics, it seems very philosophical, and the purpose of this section is to convince the reader otherwise: it is of the utmost importance to always be very sure that the question one is thinking about is actually well-posed. Some questions are not questions, they are a bunch of words arranged so convincingly that one may be misled into believing that they carry meaning! Or, as proposed by Wittgenstein (more about him in §10), *Everything that can be thought at all can be thought clearly. Everything that can be said can be said clearly* (Tractatus, 4.116). Alas, this is not always easy.

9.1 Grandi's series

We jump right in and ask a simple question: what is the value of

$$1 - 1 + 1 - 1 + 1 - 1 + 1 - 1 + 1 - \cdots \qquad ?$$

This infinite sum is known as *Grandi's series*. It is named after the Italian monk and mathematician Guido Grandi (1671–1742). He made the following claim.

Claim (Grandi) We have $1 - 1 + 1 - 1 + 1 - 1 + 1 - 1 + 1 - \cdots = 1/2$.

'Proofs'. We will give four proofs. Grandi observed

$$1 = 1$$
$$1 - 1 = 0$$
$$1 - 1 + 1 = 1$$
$$1 - 1 + 1 - 1 = 0$$

so we get 0 half of the time and 1 half of the time. This means that 'on average' we should get 1/2. Here's another proof that comes to the same conclusion. It uses the

© The Author(s), under exclusive license to Springer Nature Switzerland AG 2025
S. Steinerberger, *The Unreasonable Elegance of Mathematics*,
Springer Undergraduate Mathematics Series, https://doi.org/10.1007/978-3-032-03815-9_9

formula for the geometric series that we already discussed before and which says

$$1 + q + q^2 + q^3 + \cdots = \frac{1}{1 - q}.$$

Plugging in $q = -1$, we get that

$$1 - 1 + 1 - 1 + 1 - 1 + 1 - 1 + 1 - \cdots = \frac{1}{1 - (-1)} = \frac{1}{2}.$$

Here is a third proof. If we write parentheses like this

$$(1 - 1) + (1 - 1) + (1 - 1) + (1 - 1) + \cdots = 0$$

while, on the other hand, if we write

$$1 - (1 - 1) + (1 - 1) + (1 - 1) + \cdots = 1.$$

The truth is probably in the middle, therefore $1/2$. The last argument, again due to Grandi, is in the form of a story. Two brothers inherit a valuable gem. They agree to share possession: at the beginning of each year the brother currently owning the gem hands it over to the brother who then keeps it for the next year. If this arrangement were to go on forever, each brother would own half a gem. $\qquad\square$

This argument, historically, gave rise to a long debate involving Marchetti, Leibniz, Wolff, Riccati and many more – a veritable Who's Who of 18th century mathematics. While none of the arguments are extremely convincing, none of them seem completely false either. Daniel Bernoulli (1700–1782) gave the following cause for concern: suppose we simply add some 0's to Grandi's series. Adding 0's does not change anything, so we may as well study

$$1 + 0 - 1 + 1 + 0 - 1 + 1 + 0 - 1 + 1 + 0 - 1 + 1 - \cdots$$

Analyzing the partial sums now shows

$$1 = 1$$
$$1 + 0 = 1$$
$$1 + 0 - 1 = 0$$
$$1 + 0 - 1 + 1 = 1$$
$$1 + 0 - 1 + 1 + 0 = 1$$
$$1 + 0 - 1 + 1 + 0 - 1 = 0$$

and so on. We see that the value is now 1 for two thirds of the time and 0 for the remaining third. Adapting some of the previous arguments, this would now suggest that the value should perhaps be $2/3$. Well, that was only one of the proofs; we had three others, surely the other ones are correct, right?

9.2 The sum $1 - 2 + 3 - \cdots$

Let us now continue the fun and see whether we can get even more extreme results.

Claim
$$1 - 2 + 3 - 4 + 5 - \cdots = \frac{1}{4}.$$

This may seem a bit controversial, but was defended by one of the greatest mathematicians of all time, Leonhard Euler [81],

> Thus, when it is said that the sum of the series $1 - 2 + 3 - 4 + 5 - 6$ etc. is $1/4$, that must appear paradoxical. [...] But I have already noticed at a previous time, that it is necessary to give the word sum a more extended meaning. [...] After having established this relationship, it is no more doubtful that the sum of this series $1 - 2 + 3 - 4 + 5$ etc. is $1/4$. (Euler, 1768)

Proof. We give two arguments, the first uses a little bit of calculus (the notion of derivatives), the second one only uses Grandi's series. Start with the geometric series

$$1 + q + q^2 + q^3 + q^4 + q^5 + q^6 - \cdots = \frac{1}{1 - q}.$$

Differentiating both sides of the equation with respect to q, we get a formula that we already derived above in §8, which reads

$$1 + 2q + 3q^2 + 4q^3 + 5q^4 + \cdots = \frac{1}{(1 - q)^2}.$$

Setting $q = -1$, we get

$$1 - 2 + 3 - 4 + 5 - \cdots = \frac{1}{4}.$$

Here is a second argument: abbreviating the answer as

$$s = 1 - 2 + 3 - 4 + 5 - \ldots$$

we can regroup the terms as follows

$$\begin{aligned}
2s &= (1 - 2 + 3 - 4 + 5 - \ldots) + (1 - 2 + 3 - 4 + 5 - 6 + \ldots) \\
&= 1 + (-2 + 3 - 4 + 5 - \ldots) + 1 - 2 + (+3 - 4 + 5 - 6 + \ldots) \\
&= 0 + (-2 + 3 - 4 + 5 - \ldots) + (+3 - 4 + 5 - 6 + \ldots) \\
&= (-2 + 3) + (3 - 4) + (-4 + 5) + (5 - 6) + \ldots \\
&= 1 - 1 + 1 - 1 + \ldots
\end{aligned}$$

which is simply the Grandi series. Since we already 'know' that Grandi's series is $1/2$, we deduce that $S = 1/4$. $\qquad \square$

9.3 The sum $1 - 2 + 4 - 8 + 16 - \cdots$

We continue with another beautiful example.

Claim

$$1 - 2 + 4 - 8 + 16 - 32 + \cdots = \frac{1}{3}.$$

Taking the geometric series and replacing q by $-q$ gives

$$1 - q + q^2 - q^3 + q^4 - q^5 + q^6 - \cdots = \frac{1}{1+q}.$$

This famous equation shows up many times in the literature, for example in a letter from Christian Wolff (1679–1754) to Leibniz.

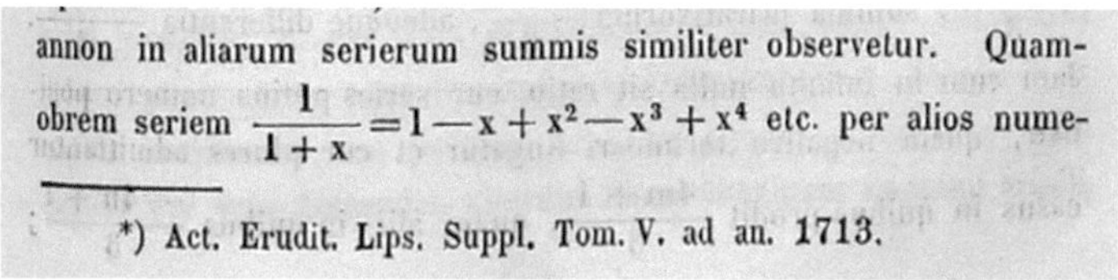

annon in aliarum serierum summis similiter observetur. Quam-
obrem seriem $\dfrac{1}{1+x} = 1 - x + x^2 - x^3 + x^4$ etc. per alios nume-

*) Act. Erudit. Lips. Suppl. Tom. V. ad an. 1713.

Fig. 9.1: The formula in a letter from Wolff to Leibniz [169].

Christian Wolff was mostly a philosopher who thought about a great many things: in 1719 he published a book with the wonderfully general title *Reasonable thoughts about God, the World, the Soul of Man, and also all other things*. Philosophers can get into trouble, and Wolff is no exception.

> In the early evening of November 12, 1723, a decree was received [...] Issued directly by Friedrich Wilhelm I, the Elector of Brandenburg and King in Prussia, the decree ordered the philosopher Christian Wolff to leave the city and all of the royal lands, within 48 hours, on pain of hanging. (Dyck [76])

One could argue that Wolff also did not shy away from mathematical hot waters: he took the equation above and plugged in all sorts of interesting values to deduce, for $q = 2$, the series claimed above

$$1 - 2 + 4 - 8 + 16 - 32 + \cdots = \frac{1}{3}.$$

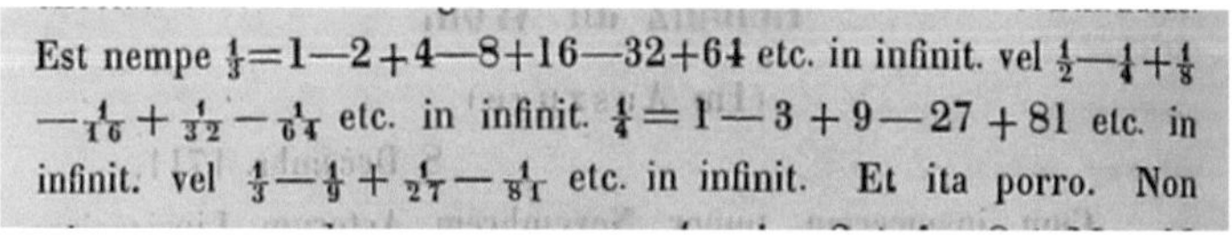

Est nempe $\tfrac{1}{3}=1-2+4-8+16-32+64$ etc. in infinit. vel $\tfrac{1}{2}-\tfrac{1}{4}+\tfrac{1}{8}$
$-\tfrac{1}{16}+\tfrac{1}{32}-\tfrac{1}{64}$ etc. in infinit. $\tfrac{1}{4}=1-3+9-27+81$ etc. in
infinit. vel $\tfrac{1}{3}-\tfrac{1}{9}+\tfrac{1}{27}-\tfrac{1}{81}$ etc. in infinit. Et ita porro. Non

Fig. 9.2: Letter from Wolff to Leibniz [169].

9.4 The sum $1 + 2 + 3 + 4 + 5 + \cdots$

We finish with one last really shocking result: even the sign feels wrong.

Claim

$$1 + 2 + 3 + 4 + 5 + 6 + \cdots = -\frac{1}{12}.$$

Proof. We are trying to determine the value of $s = 1 + 2 + 3 + 4 + 5 + 6 + \cdots$ We compare this sum with four times its value

$$s = 1 + 2 + 3 + 4 + 5 + 6 + 7 + 8 + \cdots$$
$$4s = 4 + 8 + 12 + 16 + \cdots$$

Subtracting one from the other in a funny order, one arrives at

$$s - 4s = 1 + (2 - 4) + 3 + (4 - 8) + 5 + (6 - 12) + 7 + \cdots$$

from which we deduce $-3s = 1 - 2 + 3 - 4 + 5 - 6 + \cdots$ which we already computed above. Recall that $1 - 2 + 3 - 4 + 5 - 6 + \cdots = 1/4$ and therefore $s = -1/12$. $\quad\square$

Is the sum of all positive integers truly $-1/12$? It depends a bit on who you ask. Some physicists will try to convince you that it is (see the subsequent section). The perplexed reader can take a deep breath: mathematics is fine. The fundamental problem with all these arguments, as announced above, lies in the way the question is asked. When Alexinus asks Menedemus whether *he had left off beating his father*, it is implied that Menedemus had once beaten his father. In a similar manner, asking

What is the value of the sum

$$1 - 1 + 1 - 1 + 1 - 1 + \cdots = ?$$

presupposes that the sum is well-defined, that it has a value. But why would that be the case? We argued in the previous chapter that some infinite sums were meaningful but this does not mean that all such constructs are well-defined. This is nicely summarized by the British mathematician Hardy [126], who writes

> ... but it is broadly true to say that mathematicians before Cauchy asked not 'How shall we define $1 - 1 + 1 \ldots$?' but 'What is $1 - 1 + 1 \ldots$?', and that this habit of mind led them into unnecessary perplexities and controversies which were often really verbal. (G. H. Hardy)

On a purely mathematical level, most of the mistakes come from the overly relaxed use of the geometric series

$$1 + q + q^2 + q^3 + q^4 + q^5 + \cdots = \frac{1}{1 - q}$$

which, as a formula, is only well-defined and meaningful when $-1 < q < 1$. It can be easy to forget the last part, and that is where most of the incorrect arguments came from. However, people have also been thinking about how one should make sense of these objects, and we will discuss such an approach.

9.5 Abel Summation

Euler, when writing that *it is necessary to give the word sum a more extended meaning*, proved to be quite the visionary. An example of such an extended meaning is Abel summation, named after the Norwegian mathematician Niels Henrik Abel (1802–1829). Abel, who died young, did most of his work in a very short amount of time and is a very impressive mathematician. Felix Klein (1849–1925), himself no slouch, wrote *I will not sound absurd if I compare his kind of productivity and his personality with Mozart's. Thus one might erect a monument to this divinely inspired mathematician like the one to Mozart in Vienna* [156]. Abel had the following nice idea: instead of trying to sum $a_0 + a_1 + a_2 + a_3 + a_4 + a_5 + \cdots$, we could also instead sum

$$a_0 + a_1 \cdot x + a_2 \cdot x^2 + a_3 \cdot x^3 + a_4 \cdot x^4 + a_5 \cdot x^5 + \cdots$$

and set $x = 1$, that does not change anything. Maybe setting $x = 1$ is not allowed, that is what happens with the geometric series, but maybe we can plug in values when $-1 < x < 1$. We could then see whether any problems arise if we pick x closer and closer to 1. We give two examples illustrating Abel's method. The first is our old friend, the Grandi series $1 - 1 + 1 - 1 + 1 - 1 + \cdots$ Following Abel's approach, we replace this by

$$1 - 1 \cdot x + 1 \cdot x^2 - 1 \cdot x^3 + 1 \cdot x^4 - 1 \cdot x^5 + \cdots$$

which is just a nice geometric series: if $-1 < x < 1$, then

$$1 - 1 \cdot x + 1 \cdot x^2 - 1 \cdot x^3 + 1 \cdot x^4 - 1 \cdot x^5 + \cdots = \frac{1}{1+x}.$$

We see that plugging in values of x that get closer and closer to 1 simply produces a right-hand side that gets closer and closer to $1/2$, and thus we could say that Grandi series sums to $1/2$ *in the sense of Abel*. To see the kind of problems that could arise, we could also try to sum $1 + 2 + 3 + 4 + 5 + 6 + \cdots$ which, using Abel's idea, becomes

$$1 + 2x + 3x^2 + 4x^3 + 5x^4 + 6x^5 + \cdots$$

Using the formula from above, we find that when $-1 < x < 1$

$$1 + 2x + 3x^2 + 4x^3 + 5x^4 + 6x^5 + \cdots = \frac{1}{(1-x)^2}.$$

This equation is now valid for $-1 < x < 1$. Following Abel's approach, we could now plug in values that get closer and closer to 1 and then see a problem: the right-hand side increases without bound. We do not get close to any specific number. Thus $1 + 2 + 3 + 4 + 5 + 6 + \cdots$ is *not* summable in the sense of Abel. Once one sees this idea, one can come up with many variations. In fact, there are many such summation methods: there is Hausdorff's method, Hölder summation, Ingham summability, the method of Lambert, the method of Le Roy, Mittag-Leffler summation, Riesz means and so on. They are motivated by different things and sometimes reach different results. We conclude with one recent example that was proposed by the physicist Yakov Zeldovich [309] in 1961: when given an infinite sum $a_0 + a_1 + a_2 + a_3 + a_4 + a_5 + \cdots$ one could instead try to sum

$$a_0 + a_1 e^{-x^2} + a_2 e^{-4x^2} + a_3 e^{-9x^2} + a_4 e^{-16x^2} + \cdots$$

and see whether one runs into any difficulties when x becomes smaller and smaller and smaller: if not, we can see whether we can plug $x = 0$ into the arising formula and call it a day. Grandi's series, for example, is indeed summable *in the sense of Zeldovich* and again has value $1/2$. Other summation methods exist. Returning to $1 + 2 + 3 + \cdots = -1/12$, *a pedestrian way to get the same result* is outlined by the physicist Bernardo Barbiellini-Amidei in a 1987 physics paper [22] about *The Casimir effect in conformal field theories*. One may write

$$1 + 2 + 3 + \cdots = 1e^{-\varepsilon} + 2e^{-2\varepsilon} + 3e^{-3\varepsilon} + \cdots \qquad \text{when } \varepsilon = 0.$$

Using $1 + 2x + 3x^2 + \cdots = 1/(1-x)^2$, this simplifies to

$$1e^{-\varepsilon} + 2e^{-2\varepsilon} + 3e^{-3\varepsilon} + \cdots = \frac{e^{\varepsilon}}{(e^{\varepsilon} - 1)^2} = \frac{1}{\varepsilon^2} - \frac{1}{12} + \frac{\varepsilon^2}{240} - \frac{\varepsilon^4}{6048} + \cdots$$

The first term becomes very large when ε becomes small, the other terms become small. The only term that truly remains a number is $-1/12$.

9.6 Thomson's Lamp

These ill-defined series have shown up in all sorts of curious places. There is a particular conundrum that is pointed out by the philosopher James Thomson in a 1954 paper [277]. It involves the idea of a lamp that is turned on and off in faster progression and very nicely combines the geometric series

$$1 + \frac{1}{2} + \frac{1}{4} + \frac{1}{8} + \cdots = 2$$

together with a Grandi-style argument. The full argument is as follows.

> There are certain reading-lamps that have a button in the base. If the lamp is off and you press the button the lamp goes on, and if the lamp is on and you press the button the lamp goes off. So if the lamp was originally off, and you pressed the button an odd number of times, the lamp is on, and if you pressed the button an even number of times the lamp is off. Suppose now that the lamp is off, and I succeed in pressing the button an infinite number of times, perhaps making one jab in one minute, another jab in the next half-minute, and so on, [...]. After I have completed the whole infinite sequence of jabs, i.e. at the end of the two minutes, is the lamp on or off? (Thomson, *Tasks and Super-Tasks* [277])

This is Grandi's problem in disguise, Thomson himself points this out.

What is the sum of the infinite divergent sequence

$$+1, -1, +1, \ldots?$$

Now mathematicians do say that this sequence has a sum; they say its sum is $1/2$. And this answer does not help us, since we attach no sense here to saying that the lamp if half-on. (Thomson, *Tasks and Super-Tasks* [277])

Mathematicians would usually agree that the sum is not defined; if one *had* to define it, one would have to specify the sense in which it is defined, maybe in the sense of Abel, or in the sense of Zeldovich, or maybe in some other sense? And maybe the lamp is neither on nor off but the room is half-illuminated? We leave the reader to ponder these mysteries further.

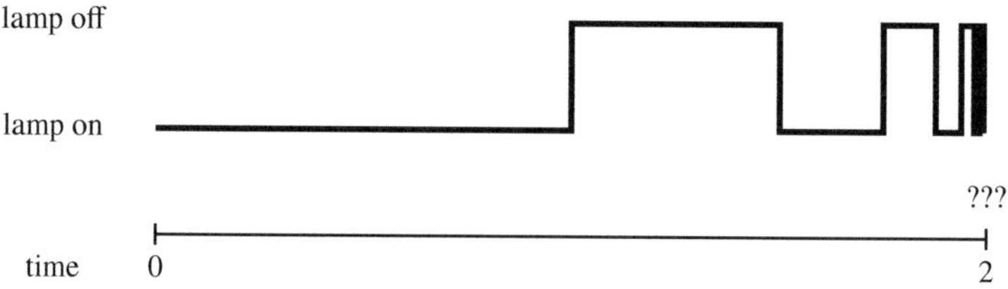

Fig. 9.3: Thomson's Lamp being turned on and off

9.7 Exercises

1. Listen to a political discussion and count how many questions are honest and how many are along the lines of *whether he had left off beating his father.*
2. Go through the proof for the geometric series formula

$$1 + q + q^2 + q^3 + \cdots = \frac{1}{1-q}$$

   and see where we use the fact that $-1 < q < 1$. The answer may be a little subtle (along the lines of 'if the thing on the left-hand is a number, then...').
3. Can you come up with other reasons (or maybe rather 'reasons') why $1 - 1 + 1 - \cdots = 1/2$?
4. Assuming the truth of Grandi's series, see whether you can come up with a convincing reason why

$$(1 - 1 + 1 - 1 + 1 - 1 + 1 - \cdots)^2 = 1 - 2 + 3 - 4 + 5 - 6 + 7 - \cdots$$

   That would be another 'proof' that the series on the right evaluates to $1/4$.
5. Get a cup of coffee, close the curtains, turn all the electronics off and contemplate Thomson's lamp. Is it on or off? What is the issue here?

Chapter 10
The Other Side of the Story (Plato vs. Wittgenstein)

10.1 Plato

We already saw some of Plato's notion of mathematics in §5. Plato thought of an abstract class of ideas that the soul has direct access to.

> **Theaetetus:** I assign them to the class of notions which the soul grasps by itself directly. (Plato, *Theaetetus*, 186a [228])

In this framework, mathematics is something that exists outside and independently of humanity. It has existed for all time and will continue to exist for all time, and we are able to access it with our soul. It serves a double purpose as being useful while also being beneficial to the soul.

> for the man of war must learn the art of number or he will not know how to array his troops, and the philosopher also, because he has to rise out of the sea of change and lay hold of true being, and therefore he must be an arithmetician. (Plato, *The Republic* [231])

The aesthetic aspect is mentioned again and again. Mathematics should not be reduced to its applicability, it deserves attention for its intrinsic beauty.

> Yes, I said, and now having spoken of it, I must add how charming the science is! and in how many ways it conduces to our desired end, if pursued in the spirit of a philosopher, and not of a shopkeeper! (Plato, *The Republic* [231])

Plato's *Republic*, discussing how one might want to organize a society, is amusing insofar as some parts of it appear to be very current: researchers in the basic science are always upset with the lack of government funding, and teachers are always upset about students not wanting to learn. This appears naturally in *The Republic* when the topic moves from planar geometry, the study of shapes in the plane, triangles, circles and so on, to the study of three-dimensional geometry.

> 'That is true, Socrates; but so little seems to be known as yet about these subjects.'
> 'Why, yes, I said, and for two reasons:—in the first place, no government patronises them; this leads to a want of energy in the pursuit of them, and they are difficult; in the second place, students cannot learn them unless they have a director. But then a director can hardly be found, and even if he could, as matters now stand, the students, who are very conceited, would not attend to him.' (Plato, The Republic [231])

There is not enough research funding, and the students are not paying attention, nothing has changed! However, Plato's view on mathematics is not the only one.

S. Steinerberger, *The Unreasonable Elegance of Mathematics*,
Springer Undergraduate Mathematics Series, https://doi.org/10.1007/978-3-032-03815-9_10

10.2 Wittgenstein

Ludwig Wittgenstein (1889–1951) is one of the most influential philosophers of the last century. As one has come to expect, his life's path is anything but straight. Born into one of the wealthiest families in Europe, he initially planned to become an engineer before turning to philosophy, where he studied with Bertrand Russell at Cambridge. He quickly made an impression.

> He maintained, for example, at one time that all existential propositions are meaningless. This was in a lecture room, and I invited him to consider the proposition: "There is no hippopotamus in this room at present". When he refused to believe this, I looked under all the desks without finding one; but he remained unconvinced. (Bertrand Russell [250])

After surviving the first World War, he published his first major work, the *Tractatus Logico-Philosophicus*. Wittgenstein was convinced to have solved *all* problems in philosophy. Wittgenstein's philosophy is not easily summarized and the same is true for the *Tractatus*. It starts in a way that is maybe reminiscent of Plato.

> *The world is all that is the case. The world is the totality of facts, not things.* (Wittgenstein, *Tractatus* [300])

Wittgenstein himself offers a one-sentence summary of the *Tractatus*: *The whole sense of the book might be summed up the following words: what can be said at all can be said clearly, and what we cannot talk about we must pass over in silence.* Wittgenstein was remarkably consistent: having solved all problems in philosophy (as he himself believed), he turned away from philosophy and became an elementary school teacher in the Austrian countryside in the early 1920s. His excessive use of corporal punishment (the 'Haidbauer incident') ended his teaching career in 1926. He then worked as gardener in a monastery and considered becoming a monk. At that point, his sister Margaret Stonborough-Wittgenstein convinced him to become an architect and design a house for her. If this seems like a strange request, we refer to Waugh's book *The House of Wittgenstein: A family at war* [295] for more details about this very unusual family. Wittgenstein became completely obsessed and spent years designing door handles, curtains, and generally agonizing over all details (frequently clashing with other people who worked on the project). The final house is a remarkable architectural construction but perhaps not the warmest of buildings. Wittgenstein's sister Hermine said *Even though I admired the house very much, I always knew that I neither wanted to, nor could, live in it myself. It seemed indeed to be much more a dwelling for the gods than for a small mortal like me* [130]. The house can still be visited in Vienna. Wittgenstein eventually returned to Philosophy and the University of Cambridge. The economist John Maynard Keynes (1883–1946) wrote to his wife

> Well, God has arrived. I met him on the 5.15 train. (Letter from John Maynard Keynes to Lydia Lopokova, January 18, 1929 [202])

In Cambridge his views on philosophy changed. There is a legend involving the Italian economist Piero Sraffa (1898–1983) and a famous Italian gesture

> [...] a conversation in which Wittgenstein insisted that a proposition and that which it describes must have the same 'logical form' [...]. To this idea Sraffa made a Neapolitan gesture of brushing his chin with his fingertips, asking: 'What is the logical form of that?' This, according to the story, broke the hold on Wittgenstein of the Tractarian idea that a proposition must be a 'picture' of the reality it describes. (Monk, [202])

One wonders how frequently an Italian hand gesture has changed the course of philosophy! We will not even attempt to hint at Wittgenstein's subsequent philosophical work ('Wittgenstein II') which is substantial, intricate, and not for the faint of heart. Instead, we will focus on a series of lectures Wittgenstein gave in 1939 that were later published under the title *Remarks on the Foundations of Mathematics*. In it, he offers a view of mathematics that is radically different from Plato. Instead of discovering mathematics, we merely play a game that we ourselves invent: at least sometimes. While the early works of Wittgenstein, notably the *Tractatus*, are filled with very definite statements, *The world is all that is the case*, the same cannot be said of his later work which is more conversational. A passage capturing some of that spirit is

> It has been said very often that mathematics is a game, to be compared with chess. In a sense this is obviously false – it is not a game in the ordinary sense. In a sense it is obviously true there is some similarity. The thing to do is not to take sides, but to investigate. It is sometimes useful to compare mathematics to a game and sometimes misleading. (Wittgenstein [303])

These lectures were attended by Alan Turing (1912–1954), a British mathematician who, just a few years later, played a pivotal role in ending World War II when he cracked the Enigma code. He was also a friend of David Champernowne, who constructed the normal number $0.12345\ldots$. Together, they designed *Turochamp*, a program designed to play Chess and maybe the first Chess program ever! Turing is also remembered as one of the first computer scientists and did pioneering work in mathematical biology. He, representing the more canonical 'Platonic' view, became one of the main discussion partners during Wittgenstein's lecture

> Turing [asked whether he understood]: I understand but I don't agree that it is simply a question of giving new meanings to words.
> Wittgenstein: Turing doesn't object to anything I say. He agrees with every word. He objects to the idea he thinks underlies it. He thinks we're undermining mathematics, introducing Bolshevism into mathematics. But not at all. We are not despising the mathematicians; we are only drawing a most important distinction between discovering something and inventing something. But mathematicians make most important discoveries. (Wittgenstein [303])

It is evident from the transcript that Wittgenstein was exceedingly fond of Turing, even going so far as to say *Unfortunately Turing will be away from the next lecture, and therefore that lecture will have to be somewhat parenthetical. For it is no good my getting the rest to agree to something that Turing would not agree to* [303]. Even though Wittgenstein ends up on the opposite side of Plato in this story, he was deeply interested in mathematics. A stunning example is *The Big Typescript* [301], a collection of 3292 pages of handwritten remarks that deal with many things but also mathematics. Section 123 deals with a proof that we already studied: Euler's proof of the infinitude of prime numbers that combines the divergence of the harmonic series with the formula for the geometric series. Wittgenstein is worried

Can we construct, from the inequality

$$1 + \frac{1}{2} + \frac{1}{3} + \frac{1}{4} + \cdots \neq \left(1 + \frac{1}{2} + \frac{1}{2^2} + \frac{1}{2^3} + \cdots\right) \cdot \left(1 + \frac{1}{3} + \frac{1}{3^2} + \cdots\right),$$

a number ν that is missing in the combinations on the right-hand side? The Eulerian proof that there are 'infinitely many prime numbers' is supposed to be an existence proof and how is such a proof possible without construction? (Wittgenstein, The Big Typescript [301])

It is worth emphasizing that these are private notes written for himself and only published after his death (*The Big Typescript* [301]). Wittgenstein continues the argument for himself and writes, somewhat poetically,

All the terms on the right-hand side occur on the left-hand side but the sum on the left side yields ∞ and the one on the right only a finite value – THEREFORE IT MUST BE ... but in mathematics nothing MUST BE, only that what IS. [...] There are no symptoms in mathematics, symptoms can only occur psychologically within the mathematician. One could also say: one cannot argue in mathematics for something, that one cannot SEE. (Wittgenstein, [301])

Much has been written about Wittgenstein's views on mathematics. Most mathematicians are Platonists, either by direct admission or by the way they approach the subject and, as such, Plato continues to have a bigger impact on mathematics; but he is not the only game in town.

10.3 Constructivism

What does it really mean to prove that a mathematical object exists? Take the statement: there exist at least 4 different prime numbers. One could argue in a direct, constructivist approach and say $2, 3, 5$ and 7 are 4 different prime numbers. One could also argue that if only three or less prime numbers existed, then

$$1 + \frac{1}{2} + \frac{1}{3} + \cdots \leq \left(1 + \frac{1}{2} + \frac{1}{2^2} + \cdots\right)^3 = 8,$$

which is a contradiction. This contradiction shows that at least 4 different prime numbers exist. These are clearly very different types of arguments: the first constructs the answer, the second shows that negating the statement contradicts another known statement. The validity of the second type of argument was once heavily debated.

Kronecker insisted that there could be no existence without construction. For him, as for Gordan, Hilbert's proof of the finiteness of the basis of the invariant system was simply not mathematics. Hilbert, on the other hand, throughout his life was to insist that if one can prove that the attributes assigned to a concept will never lead to a contradiction, the mathematical existence of the concept is thereby established. (Reid, [241])

One of the main protagonists in this story is Luitzen Egbertus Jan Brouwer (1881–1966). After doing brilliant work in his youth, he started to doubt the *law of excluded middle*, the idea that a statement must be either true or false. This notion is a key ingredient in non-constructive proofs: if the statement that there are at most 3 different prime numbers is wrong, then its negation, there are at least 4 different prime numbers, must be true – but only if it is indeed the case that for each statement P and its negation not P exactly one of them is true and one of them is false. Brouwer argues that this type of idea, the law of excluded middle, cannot be tested

> An a priori character was so consistently ascribed to the laws of theoretical logic that until recently these laws, including the principle of excluded middle, were applied without reservation even in the mathematics of infinite systems and we did not allow ourselves to be disturbed by the consideration that the results obtained in this way are in general not open, either practically or theoretically, to any empirical corroboration. (Brouwer [129])

Brouwer draws the following striking analogy *an incorrect theory, even if it cannot be inhibited by any contradiction that would refute it, is none the less incorrect, just as a criminal policy is none the less criminal even if it cannot be inhibited by any court that would curb it* [129]. Wittgenstein, in his *Big Typescript*, approves: *When Brouwer battles against the application of the law of the excluded middle in mathematics, he is right in so far as he is directing his attack against a process that is analogous to the proofs of empirical propositions. In mathematics you can never prove anything this way: I saw two apples lying on the table, and now there is only one there, so A has eaten an apple* [301]. Constructivist concerns are not talked about as much these days. It is not that these questions are not real questions, they are, but the mathematics that one is able to do without the law of excluded middle turns out to be a little bit boring: it is harder to prove things and one is not able to prove as many things as one would like. Since there are no definite ways of resolving such philosophical concerns in any case, people prefer to go with the type of mathematics that is the most fun – and that includes the law of excluded middle.

10.4 Finitism

Speaking of philosophy and different points of view on mathematics, we quickly also mention finitism (which is not too far from Wittgenstein's position) and ultrafinitism. While Plato and most of us are not particularly troubled by the integers $1, 2, 3, 4, \ldots$, one may naturally wonder whether the use of dots $(\ldots)$ is legitimate. Finitists accept the existence of individual integers such as 3 or 18 but draw the line at *infinite* sets, they do not allow the use of the dots $(\ldots)$. Some even go as far as to express concern about *very* large numbers, sometimes in a humorous way

> I have seen some ultrafinitists go so far as to challenge the existence of 2^{100} as a natural number, [...] I raised just this objection with the (extreme) ultrafinitist Yessenin-Volpin during a lecture of his. He asked me to be more specific. I then proceeded to start with 2^1 and asked him whether this is "real" or something to that effect. He virtually immediately said yes. Then I asked about 2^2, and he again said yes, but with a perceptible delay. Then 2^3, and yes, but with more delay. This continued for a couple of more times, till it was obvious how he was handling this objection. Sure, he was prepared to always answer yes, but he was going to take 2^{100} times as long to answer yes to 2^{100} then he would to answering 2^1. There is no way that I could get very far with this. (Friedman [93])

Considering the number of different ideas of what mathematics is, it feels important to encourage people to pursue their own ideas, even or especially if they are perhaps a little bit unconventional. In this spirit, it feels appropriate to conclude with a description that Brouwer gave in a 1906 letter to his friend Adama van Scheltema: *Life is a magic garden. With wondrous softly shining flowers, but between the flowers there are the little gnomes, they frighten me so much, they stand on their heads, and the worst is, they call out to me that I should also stand on my head, every once in a while I try, and I die of embarrassment; but sometimes the gnomes shout that I am doing very well, and that I'm indeed a real gnome myself after all. But on no account I will ever fall for that* [40].

10.5 Exercises

1. Learn a bit more about Wittgenstein's life. His older brother Paul Wittgenstein (1887–1961) was a pianist who lost his arm in World War 1 and became a one-armed pianist. He convinced composers to write music for him, including Maurice Ravel (Piano Concerto for the Left Hand) and Sergei Prokofiev (Piano Concerto No. 4 in B-flat major for the left hand, Op. 53).
2. Wittgenstein had significant interactions with Frank P. Ramsey (1903–1930), who made significant advances in philosophy ('Ramsey sentences', 'The Ramsey test'), mathematics ('Ramsey theory') and economics ('Keynes–Ramsey rule', 'Ramsey–Cass–Koopmans model'), before dying at the age of 26. Read a little bit about Ramsey's life and ponder how much time we constantly waste.
3. Does 'mathematics' exist? Is it as real as the room you are currently in?
4. Compare the constructive proof of '4 different prime numbers exist' to the non-constructive proof given above. Are both proofs equally valid? Try to find arguments for both sides.
5. The British philosopher G. E. Moore argues in *Proof of an External World* that

   I can prove now, for instance, that two human hands exist. How? By holding up my two hands, and saying, as I make a certain gesture with the right hand, 'Here is one hand', and adding, as I make a certain gesture with the left, 'and here is another'. [204]

   Do you find it convincing? If not, why not? There is a noticeable historical precedent recounted in Boswell's Biography of Samuel Johnson *After we came out of the church, we stood talking for some time together of Bishop Berkeley's ingenious sophistry to prove the non-existence of matter, and that every thing in the universe is merely ideal. I observed, that though we are satisfied his doctrine is not true, it is impossible to refute it. I never shall forget the alacrity with which Johnson answered, striking his foot with mighty force against a large stone, till he rebounded from it, 'I refute it **thus**.'* [33].

Chapter 11
Different Infinities

there is [...] a multitude which no man can number,
and they cause me endless trouble. (Plato, [230])

11.1 'Infinity' as 'very large'

Suppose we have two different sets, these are collections of objects, and we are trying
to understand whether they have the same number of elements. When the number of
elements in each set is finite, for example the sets

$$\{a, b, c, d, e\} \qquad \text{and} \qquad \{1, 2, 3, 4, 5\},$$

then this is simple: we count the number of elements in each set and see whether
we get the same number twice. There are 5 objects in $\{a, b, c, d, e\}$ and 5 objects,
conveniently self-enumerating, in $\{1, 2, 3, 4, 5\}$. Since $5 = 5$, clearly these two sets
have the same number of elements. When one of the sets is infinite, like $\mathbb{N} =
\{1, 2, 3, \dots\}$ and the other set is finite, like $\{a, b, c, d, e\}$, then this is also easy to
answer: they have a different number of elements because 5 is smaller than infinity.
Now we ask a more challenging question: what if both sets have an infinite number
of elements, for example $\mathbb{N} = \{1, 2, 3, \dots\}$ and the set of real numbers $\mathbb{R}$? At first
glance, there is an easy answer.

> **Easy answer.** If two sets have an infinite number of elements, then they have the same
> number of elements because 'infinity is infinity'.

This was how people dealt with the issue for a long time – and it's not necessarily a
bad way of dealing with it. One should observe that, frequently, people use 'infinite'
merely as a placeholder for 'very large'. A nice example is found in *The Sand-
Reckoner* [7] by Archimedes, written more than 2000 years ago:

> There are some, King Gelon, who think that the number of the sand is infinite in multitude;
> and I mean by the sand not only that which exists about Syracuse and the rest of Sicily but
> also that which is found in every region whether inhabited or uninhabited. Again there are
> some who, without regarding it as infinite, yet think that no number has been named which
> is great enough to exceed its multitude. (Beginning of Archimedes' *The Sand-Reckoner* [7])

Archimedes proceeds to construct a number larger than that of *a mass [of sand] equal
in magnitude to the Universe* [7]. It is interesting to read Archimedes' definition of
what he means by 'universe': *'Universe' is the name given by most astronomers to
the sphere whose centre is the centre of the Earth and whose radius is equal to the*

© The Author(s), under exclusive license to Springer Nature Switzerland AG 2025   107
S. Steinerberger, *The Unreasonable Elegance of Mathematics*,
Springer Undergraduate Mathematics Series, https://doi.org/10.1007/978-3-032-03815-9_11

straight line between the centre of the sun and the centre of the Earth. Back then, the Earth was still in the center and the Sun revolved around the Earth. However, Archimedes remarks, more radical notions also exist.

> But Aristarchus of Samos brought out a book consisting of some hypotheses, in which the premisses [sic!] lead to the result that the universe is many times greater than that now so called. His hypotheses are that the fixed stars and the sun remained unmoved, that the earth revolves about the sun in the circumference of a circle, the sun lying in the middle of the orbit [...] (Archimedes, [7])

Archimedes then proceeds to make some basic assumptions about the size of the universe and argues that even if it were gigantic, the amount of sand in it would still be finite, not infinite! 'Infinite' is often just a placeholder for 'very large'. Some things are actually infinite: for example, as we have seen above, there are infinitely many prime numbers. There are more prime numbers than there is sand in the universe (even if the universe was *very* large and filled entirely with sand). However, even here there is a trick to circumvent the use of 'infinity'. If someone says that 'there are infinitely many prime numbers' one could rephrase the sentence as saying 'for any number N, it is possible to find N different prime numbers' and now the statement no longer contains any notion of 'infinity', it is a purely finite statement for every single $N \in \mathbb{N}$. Adopting Wittgenstein's maxim *Whereof one cannot speak, thereof one must be silent* (Wittgenstein, [300]), we may be content to speak of finite numbers and of nothing else; and maybe 'infinite' is simply a placeholder for that one cannot speak of. Not a bad way to live and well aligned with the Finitists we met in the previous chapter.

11.2 Different Infinities

Thinking of 'infinity' as a black box for 'very large' or, maybe, as a polite term of 'that which shall not be discussed' is one way of dealing with the issue – but it is not the only one. As early as 1225, Robert Grosseteste writes in his *De Luce* (On Light) that not only does infinity exist, there is more than one such notion of infinity,

> And some infinites are larger than other infinites, and some are smaller. (Grosseteste, [117])

This is indeed what has come to pass as the standard point of view in mathematics today and we will spend the remainder of the chapter exploring why that is the case. The original argument of Grosseteste is amusing but does not hold up to modern scrutiny. It goes as follows: *Thus the sum of all numbers both even and odd is infinite. It is at the same time greater than the sum of all the even numbers although this is likewise infinite, for it exceeds it by the sum of all the odd numbers* [117]. Writing this out formally, Grosseteste says that $A = 1 + 2 + 3 + \cdots$ is surely infinite (which is true) and that

$$B = 2 + 4 + 6 + 8 + 10 + \cdots$$
$$C = 1 + 3 + 5 + 7 + 11 + \cdots$$

are also each infinite (also true) and that $A = B + C$ and thus $A > B$. The last two steps of reasoning may remind the reader of the arguments with divergent series: they look alright but they seem to lack substance. Indeed, we can go one step further and produce a contradiction: naively subtracting A from B shows that

$$(B - A) = (2 + 4 + 6 + 8 + 10 + \cdots) - (1 + 2 + 3 + 4 + 5 + \cdots)$$
$$= (2 - 1) + (4 - 2) + (6 - 3) + (8 - 4) + \cdots = 1 + 2 + 3 + \cdots = A,$$

which shows that $B - A = A$ and suddenly B is twice as large as A. Robert Grosseteste was a statesman, scholastic philosopher, and the Bishop of Lincoln in England. For some time after his death, he was even revered as a saint, however, the formal process of making him a saint seems to have failed for an interesting reason: his *ghost* was accused of killing the pope. Grosseteste had what one might call philosophical differences with Pope Innocent IV. This is understandable: Innocent IV authorized torture against heretics *provided he does so without killing them* [145].

The end of the story is intriguing and told by Matthew of Paris in his *Chronica Majora*: in this book Matthew describes all of history starting from Creation to 1259 (the year of his death). A year after Robert Grosseteste's death in 1253, *One day, in the same year, the pope, in an excessive fit of anger, wished [...] to throw the bones of Robert, Bishop of Lincoln, out of the church, and to hurl them to such infamy and degradation that he might be proclaimed a heathen and a disobedient rebel throughout the world [...] But in the night following this day, a vision appeared to the Pope* [221]. This vision is of none other than Robert Grosseteste who, wearing formal Bishop's gown, said with a terrible voice *Miserable Pope, did you propose to cast out my bones from the Church of Lincoln?* The Ghost then *stabs him in the side with an impetuous blow with the Bishop's staff* and disappears *leaving the Pope half-dead [...] His chamberlains were astonished at hearing his exclamations [...]* he replied 'The terrors of the night have greatly disturbed me, and I shall never be restored to my former health.' Indeed, Pope Innocent IV died shortly thereafter.

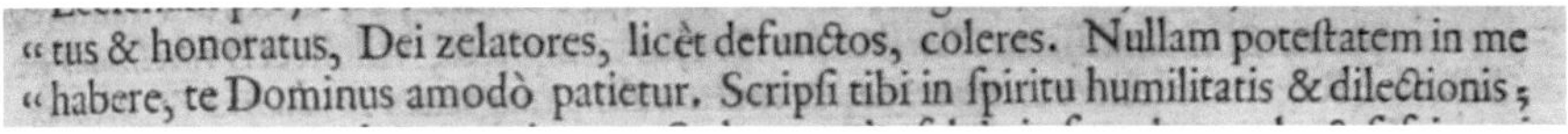

« tus & honoratus, Dei zelatores, licèt defunctos, coleres. Nullam potestatem in me
« habere, te Dominus amodò patietur. Scripfi tibi in fpiritu humilitatis & dilectionis ;

Fig. 11.1: The Ghost of Robert Grosseteste telling Pope Innocentius IV: You have no power over me, *the Lord will not bear it* as told in the *Chronica Majora* [221].

11.3 The John Duns Scotus and Galileo mysteries

Infinity presents its own challenges, and we will discuss two of them. Around the time of Robert Grosseteste, John Duns Scotus (literally *John of Duns, the Scotsman*) (1266–1308), whose nickname was *Doctor Subtilis*, asks two seemingly innocent questions (see Fig. 11.2). First, Duns Scotus says, consider a square and the diagonal inscribed in the square. Clearly, the diagonal is longer than one side of the square (and we even know how much longer: by a factor of $\sqrt{2}$). However, we can draw these straight parallel lines and identify each point on the diagonal with exactly one point on the left side-length. Therefore there are as many points on the side as there are on the diagonal. *This is impossible*, Scotus, declares.

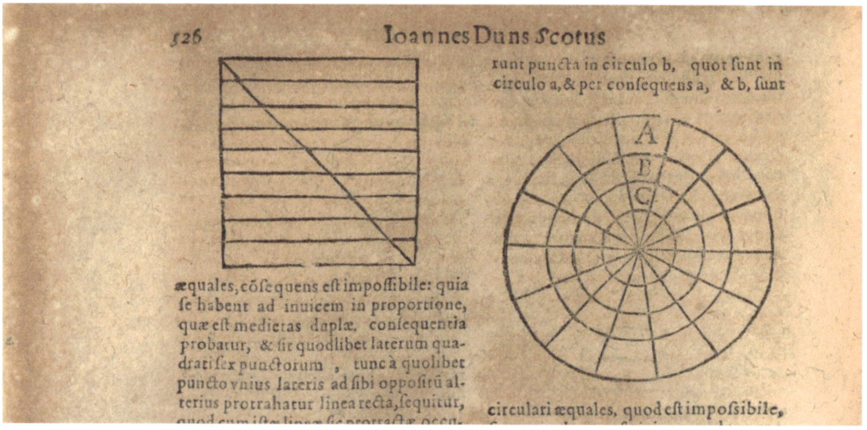

Fig. 11.2: The Geometric Challenges of John Duns Scotus [258]

However, the argument is indeed correct: for each point on the side of the square, we can find a unique point on the diagonal and, conversely, for each point on the diagonal, we find a unique point on the side. The number of points match. Or, Duns Scotus continues, consider three circles of different radii centered at the same point (see Fig. 11.2, right). Again, we can identify each point on the outermost circle with a unique point on the innermost circle. *Then there will be as many points in the circle B as there are in the circle A [...] which is impossible* [258]. John Duns Scotus, the *Doctor Subtilis*, had an unusual death: being afflicted with catalepsy, a neurological disease characterized by fixture of posture and unresponsiveness to external stimuli, he was one day found in his chambers, thought to be dead and buried alive – or at least so it is reported by Francis Bacon in his *History of Life and Death*. His posthumous reception is also most unusual.

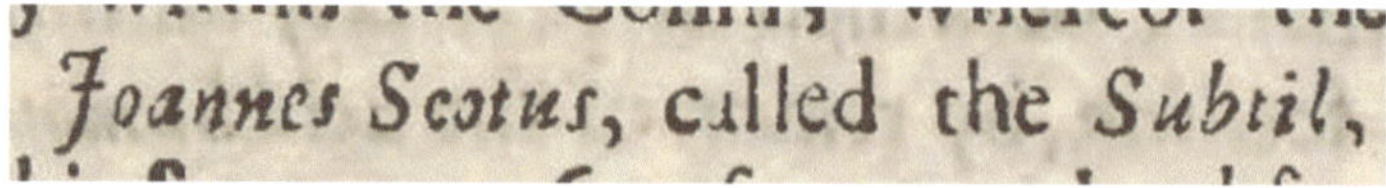

Fig. 11.3: Left: Francis Bacons's *History of Life and Death* [16] describing *the most recent and memorable example was that of Joannes Scotus, called the Subtil, and a School-man*. Bacon explains that Scotus was mistakenly buried alive.

At first, he was widely respected for his works ('Scotism'). The Italian polymath Gerolamo Cardano [47] names him as one of the 12 greatest scholars of all time, and his followers are referred to as 'Dunces', the ones following Duns Scotus. Then, something changed, and Duns Scotus is now rarely even mentioned – the change is most noticeably in the meaning of 'dunce' which now commonly means something like 'idiot' or 'dullard'. Some people may remember the 'dunce cap', a cone-shaped hat that is placed on particularly unstudious school children (a practice less common today). Hobbes, when attacking Wallis in his *Marks of the absurd geometry etc.*

> As for the judgment of that public professor that makes himself a witness of the goodness of your geometry, a man may easily see by the letter itself that he is a dunce. (Hobbes [136])

In what must be a singular event, John Duns Scotus, the priest, Franciscan Friar and supreme theologian, *the Subtle One*, had such an impact that 700 years later his name appears as an *insult* in a contemporary translation of the Bible

> Even *dunces* who keep quiet are thought to be wise; as long as they keep their mouths shut, they're smart. (Proverbs 17:28 in the MSG translation)

John Duns Scotus and his subtle geometric puzzles aside, another person who struggled with infinite sets and their consequences was Galileo Galilei (1564–1642). In his 1638 *Discourses and Mathematical Demonstrations Relating to Two New Sciences*, Galileo noted the following imagined conversation between Simplicio and Salviati, which deserves to be read out aloud slowly.

> **Salviati.** This is one of the difficulties which arise when we attempt, with our finite minds, to discuss the infinite, assigning to it those properties which we give to the finite and limited; but this I think is wrong, for we cannot speak of infinite quantities as being the one greater or less than or equal to another. To prove this I have in mind an argument which, for the sake of clearness, I shall put in the form of questions to Simplicio who raised this difficulty. I take it for granted that you know which of the numbers are squares and which are not. [...] Therefore if I assert that all numbers, including both squares and non-squares, are more than the squares alone, I shall speak the truth, shall I not?
> **Simplicio.** Most certainly.
> **Salviati.** If I should ask further how many squares there are one might reply truly that there are as many as the corresponding number of roots, since every square has its own root and every root its own square, while no square has more than one root and no root more than one square.
> **Simplicio.** Precisely so.
> **Salviati.** But if I inquire how many roots there are, it cannot be denied that there are as many as there are numbers because every number is a root of some square. This being granted we must say that there are as many squares as there are numbers because they are just as numerous as their roots, and all the numbers are roots. (Galileo, [96])

Salviati here claims that the set of all positive integers $\mathbb{N} = \{1, 2, 3, 4, 5, 6, \dots\}$ is surely larger than the set of squares of positive integers $\{1, 4, 9, 16, 25, 36, \dots\}$. This is easy to see: each square number, like 25 or 49, is an integer but there are many other integers that are not square numbers. So this tells us that there are more positive integers than there are squares of positive integers. However, Salviati proceeds, we can also identify each square number with its root

$$1 \leftrightarrow 1, \quad 4 \leftrightarrow 2, \quad 9 \leftrightarrow 3, \quad 16 \leftrightarrow 4, \quad 25 \leftrightarrow 5 \dots$$

This one-to-one identification suggests that the number of squares of integers and the number of positive integers is really the same – but that contradicts the fact that there are more integers than squares. How should one resolve this issue? Salviati replies *So far as I see we can only infer that the totality of all numbers is infinite, that the number of squares is infinite, and that the number of their roots is infinite; neither is the number of squares less than the totality of all numbers, nor the latter greater than the former; and finally the attributes "equal," "greater," and "less," are not applicable to infinite, but only to finite, quantities* [96]. This mystery is similar to the geometric John Duns Scotus problem (two circles, one clearly being larger, and yet there being a one-to-one correspondence between their points suggesting their number to be the same). Galileo argues in favor of abandoning the idea of the 'size' of an infinite set. Infinite is infinite and that's that.

11.4 Cantor's Idea

The story continues with the German mathematician Georg Cantor (1845–1918) and culminates in modern set theory. Modern set theory agrees with Grosseteste (*And some infinites are larger than other infinites, and some are smaller*). Cantor's idea is very natural: given two sets A and B, finite or infinite, when is it appropriate to say they have the same number of elements? To understand Cantor's idea, we will need bijective functions. A function $f : A \to B$ is bijective if it satisfies two properties

1. if $a_1 \neq a_2 \in A$ are different, then $f(a_1) \neq f(a_2)$ are also different
2. for every $b \in B$ there exists an $a \in A$ such that $f(a) = b$.

This may sound a little bit abstract, so let us look at an example. When $A = \{a, b, c, d, e\}$ and $B = \{1, 2, 3, 4, 5\}$, then an explicit function could be

$$f(a) = 1, f(b) = 2, f(c) = 3, f(d) = 4, f(e) = 5.$$

No two elements from A get sent to the same element of B and each element in B has an element from A assigned to it. Instead of talking about *size* or *number of elements*, we will instead talk about the *cardinality*, which is a new idea.

Definition (Georg Cantor). Two sets A and B are said to have the same cardinality if there exists a bijective function $f : A \to B$.

It is not too difficult to see that this notion is symmetric: if there exists a bijective function $f : A \to B$, then there are also exists a bijective function $g : B \to A$. We already encountered such bijective functions before: Duns Scotus draws straight lines and sends each point on the side to a unique point on the diagonal (that is not hit by any other line). Likewise, in the example of the concentric circles, he is mapping points radially to other points. Duns Scotus is constructing bijections! Likewise, Galileo describes an explicit projection $f : \mathbb{N} \to \{\text{squares}\}$ by writing $f(n) = n^2$. His argument would today perhaps look a little bit like this.

Theorem. *The set* $\mathbb{N} = \{1, 2, \dots, \}$ *and the set of squares* $\{1, 4, 9, 16, \dots \}$ *have the same cardinality.*

Proof. If we want to claim that two sets have the same cardinality, we have to find a bijective function f sending one set to the other set. Here, the function $f : \mathbb{N} \to \{1, 4, 9, 16, \dots\}$ is given by $f(n) = n^2$. Two different positive numbers $m \neq n$ get sent to different square numbers because, if m, n are positive, then

$$m \neq n \quad \text{implies that} \quad m^2 \neq n^2.$$

Moreover, each square number n^2 is reached by some number since $n^2 = f(n)$. $\quad\square$

If we think of A, B having the same cardinality as a way of saying that they have the same size, this may be troubling: the set of square numbers $\{1, 4, 9, 16, \dots\}$ is a *strict* subset of the integers $\mathbb{N}$. One could also say, as Galileo did, that $\mathbb{N}$ contains all the square numbers *and then also many other numbers*. So surely it would not make sense to say that they are the same size, clearly $\mathbb{N}$ should be much larger. But maybe it is okay for them to have the same *cardinality*? A new word can be most invigorating! One could produce other shocking examples. Here, for example, is a simple proof that the set of all integers (positive or negative)

$$\mathbb{Z} = \{\dots, -2, -1, 0, 1, 2, \dots\} \quad \text{and} \quad \mathbb{N} = \{1, 2, \dots\}$$

have the same cardinality. We will do this by producing a completely explicit bijective map $f : \mathbb{N} \to \mathbb{Z}$ that sends nonnegative integers to all integers.

positive integer n	1	2	3	4	5	6	7	8	$\cdots$
integer $f(n)$	0	-1	1	-2	2	-3	3	-4	$\cdots$

Fig. 11.4: How to build a bijective $f : \mathbb{N} \to \mathbb{Z}$.

This shows that $\mathbb{N}$ and $\mathbb{Z}$ have the same cardinality, which is at the very least counterintuitive: one would expect $\mathbb{Z}$ to be twice as large as $\mathbb{N}$. It gets worse: if we compare $\mathbb{N}$ with all points in $\mathbb{R}^2$ with integer coordinates such as $(0, 0)$ and $(-4, 2)$, a set denoted by $\mathbb{Z}^2$, then the second set should be *infinitely* larger than the first.

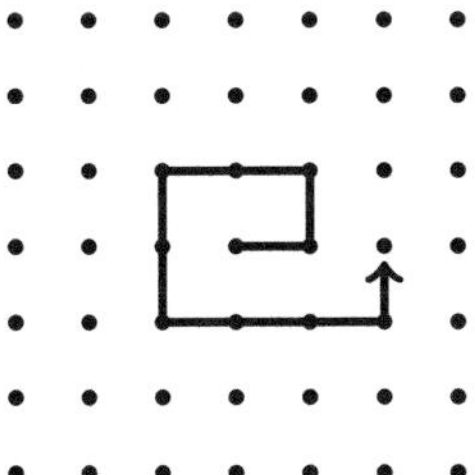

Fig. 11.5: Creating a one-to-one map between $\mathbb{N}$ and all the points in the plane with integer coordinates by moving in a snake-like fashion.

Elements in $\mathbb{N}$ can be described with one number, elements in $\mathbb{Z}^2$ are made up of *two* numbers. $\mathbb{Z}^2$ should be 'two-dimensional' while $\mathbb{N}$ is 'one-dimensional'. From the perspective of cardinality, the sets $\mathbb{N}$ and $\mathbb{Z}^2$ are of the same size! Fig. 11.5 shows how this can be done, we move around in a snake-like fashion and enumerate $f(1) = (0,0), f(2) = (1,0), f(3) = (1,1), f(4) = (0,1), f(5) = (-1,1)$ and so on. The snake will eventually eat all the points!

11.5 The Cantor controversy

The reaction of the mathematical 'establishment' to Cantor's ideas is a rich and complicated story. It is remarkable that, ultimately, everything boils down to a question of definitions and axioms and verifiable arguments – it is not about discussing the relative merits of a poem or deciding which musical composition deserves to win an award. On the other hand, evaluating mathematics is perhaps not all that different from appreciating art. The statements *There are infinitely many prime numbers* and $2412 + 1000 = 3412$ are equally true but it does appear that the first statement feels more... *substantial?* This is clearly starting to be a matter of taste. If extraterrestrials were to land tomorrow and we presented them with both statements, it is not at all clear what their reaction would be – assuming they are able to build a spaceship, they presumably have developed enough mathematics to be familiar with both statements, but maybe they feel that prime numbers are kind of ugly, and maybe they really, really like a good old-fashioned addition.

Fig. 11.6: Georg Cantor, Professor at the University Halle-Wittenberg.

All of this is nicely illustrated by how Cantor's ideas were perceived. It is a wonderful example of how mathematics, like everything else, is governed by passions, trends, hypes and fads. Humanity is undertaking the study of eternal truths in a characteristically human fashion. The story is perhaps even more surprising if we emphasize that Cantor's work is now considered a completely standard branch of mathematics and very frequently taught at the undergraduate level.

> Cantor was then in his fortieth year. Over 10 years had passed since the first appearance of his fundamental works. With Kronecker, he found the sharpest opposition. It does not exceed the permissible limit when I say that Kronecker's attitude must have given the impression as if Cantor, in his capacity as a researcher and teacher, was a corrupter of youth. Although he had a ready place for his publications in the Annalen, scientific influence on others or even visible recognition hardly accrued to him in Germany before the year 1884. He felt lonely and abandoned, even by those whom he revered. (Schoenflies [255])

Kronecker exerted a lot of scientific influence, and Cantor felt sidelined. The frustration is evident in a letter of his to the Swedish mathematician Mittag-Leffler where Cantor is appealing to Vergil's phrase 'I fear the Greeks, even when they are bringing presents' (referring to the Trojan horse).

> My dear friend! [...] I would like to warn you not to readily consider the publication intended by Kronecker as worthy of inclusion in your journal. In dealing with this man, keep in mind the words 'timeo Danaos et dona ferentes'. (Cantor to Mittag-Leffler, 25. Jan. 1884, [255])

A little bit later, in 1885, Cantor tried to publish two short articles in *Acta Mathematica*, the journal edited by Mittag-Leffler. Mittag-Leffler replied to his friend encouraging Cantor *not* to publish

> I am convinced that the publication of your new work [...] will greatly damage your reputation among mathematicians. I know very well that basically this is all the same to you. But if your theory is once discredited in this way, it will be a long time before it will again command the attention of the mathematical world. It may well be that you and your theory will never be given the justice you deserve in your lifetime. (Mittag-Leffler to Cantor, 9. Mar. 1885, [70])

Cantor was devastated. His friendship with Mittag-Leffler cooled, and he never tried to publish in *Acta Mathematica* again. Discouraged and isolated, he expressed *little interest in my mathematical research* [70] and turned to other pursuits. *Between 1885 and 1891 what material he did publish appeared in journals devoted to philosophy, and what survives from this period of his personal correspondence reveals a sustained preoccupation with the Bacon–Shakespeare controversy, Rosicrucianism, Freemasonry, and various historical-literary pursuits* [70]. The Bacon–Shakespeare controversy continued to occupy Cantor far beyond this period: it deals with the question of whether William Shakespeare (about whom, as a person, we know surprisingly little) truly existed or whether his works may have been written by someone else – the most popular candidates are Francis Bacon, Edward de Vere, and Christopher Marlowe. Cantor himself published on the subject. In 1899, he wrote a letter to the Ministry of Culture in which he writes

> I want to add that in the last fifteen years I completely resolved the question of the authorship of the works of Shakespeare in favor of Bacon of Verulam, the Vice-Count of St. Albans (Francis Bacon) and I have also come to historical insights about the first kings of Great Britain which will undoubtedly shake the English Government as soon as they are published (Letter from Cantor to Count Posadowsky-Wehner, 10. Nov. 1899 [116])

Doubting the existence of Shakespeare is not as far-fetched as it may seem, Cantor is in good company. Sigmund Freud had his own doubts and even suggested that the name may be a mispronunciation of the French name *Jacques Pierre*. He later wrote that *I no longer believe in the man from Stratford* and arguing in favor of *the noble, highborn, and finely cultured, the passionately disordered, the somewhat declassé aristocrat Edward de Vere* [139].

11.6 Different Infinities

In 1891 and despite all the opposition, Georg Cantor published what is now known as *Cantor's Diagonal Argument* [43]. It is one of the truly amazing eternal ideas in mathematics. Cantor uses it to prove that there are two different infinite sets that do *not* have the same cardinality – he is proving what was already stated by Grosseteste 600 years earlier *And some infinites are larger than other infinites, and some are smaller.* There are, Cantor shows, more real numbers in the unit interval $[0, 1]$ than there are positive integers: these sets do not have the same cardinality.

Theorem (Cantor). *The sets $\mathbb{N}$ and $\mathbb{R} \cap [0, 1]$ do not have the same cardinality.*

Proof. We have to show that there cannot be a bijective map

$$f : \mathbb{N} \to \mathbb{R} \cap [0, 1].$$

The proof is a proof by contradiction. We assume such a map exists. If that is the case, then this would correspond to an *enumeration* of the real numbers in $[0, 1]$. The number $f(1)$ would correspond to the first element in the sequence of real numbers, $f(2)$ to the second element and so on. Well, we can assume (for the sake of contradiction) that we have such an enumeration

$$f(1) = 0.a_{11}a_{12}a_{13}a_{14}\ldots$$
$$f(2) = 0.a_{21}a_{22}a_{23}a_{24}\ldots$$
$$f(3) = 0.a_{31}a_{32}a_{33}a_{34}\ldots$$
$$\ldots$$

where a_{ij} is the j-th digit in the decimal expansion of the real number $f(i)$. Cantor's Diagonal Argument will now show that for each such list, there exists a real number in $[0, 1]$ that is *not* contained in the list: that violates the bijectivity of f. Cantor's argument is fantastic: once understood, it becomes unforgettable. One simply *builds* a real number that is missing from the list. The number has to be in $[0, 1]$, so it will start with a 0. For the first digit after the decimal point, we use any digit that is different from a_{11}. For the second digit after the decimal point, we can use any digit except a_{22}. For the third digit from the decimal point we use any digit except a_{33}. If we continue this for all time, then the final number that we get is guaranteed to not be on the list. To make this concrete, suppose our list starts with

$$f(1) = 0.123456789\ldots$$
$$f(2) = 0.234567890\ldots$$
$$f(3) = 0.345678901\ldots$$
$$f(4) = 0.456789012\ldots$$

What would our hypothetical number look like? It starts with 0 and for the first digit after the decimal point, we can use any digit other than 1, maybe 5, that would give us 0.5. For the second digit after the decimal point, we can use any digit other than 3, so maybe 1, giving us 0.51. For the third digit after the decimal point, we can use any digit other than 5, so maybe 8. This gives us 0.518 and this construction can be continued indefinitely to give us a number

$$x = 0.518\ldots$$

We now claim that this number is not on the list. It cannot be $f(1)$ because the first digit of $f(1)$ after the decimal point is 1 and we deliberately chose the first digit of x to be different from 1. It also cannot be $f(2)$ or $f(3)$ for the same reason (the second digit is different from the second digit of $f(2)$ and likewise for the third digit and $f(3)$). Could it be $f(121)$? It cannot be, because the 121th digit of x was chosen to be different from the 121th digit of the 121th number, $f(121)$. So $x \neq f(121)$. This argument always works, and therefore x is not on the list. $\quad\square$

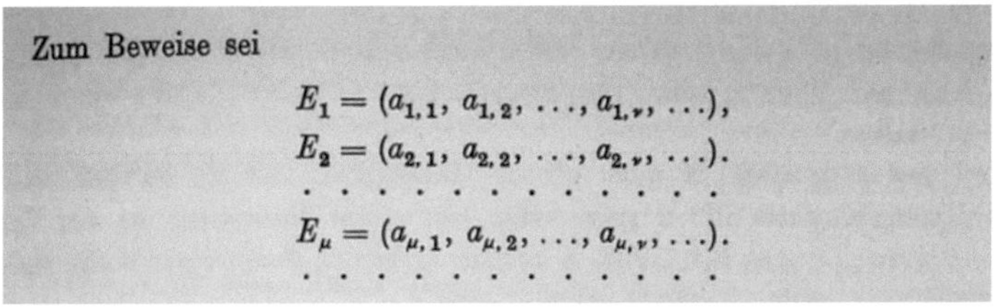

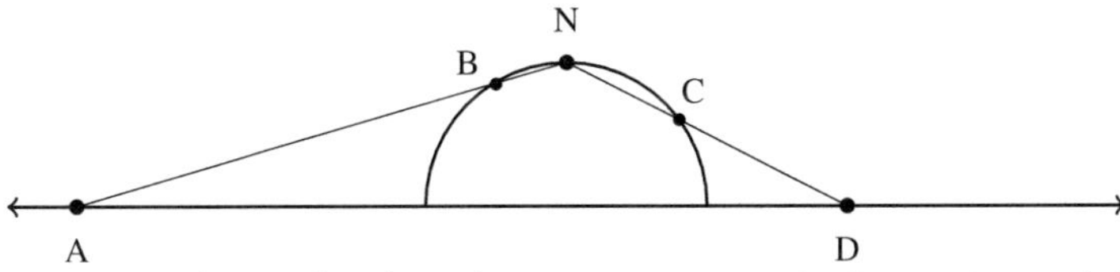

Fig. 11.7: *For the proof, let...* Cantor's diagonal argument [43].

This shows that $\mathbb{N}$ and $\mathbb{R}\cap[0,1]$ do not have the same cardinality. It is counterintuitive but not very difficult to show that $\mathbb{R}\cap[0,1]$ and $\mathbb{R}$ have the same cardinality. This then implies that $\mathbb{N}$ and $\mathbb{R}$ have a different cardinality.

Theorem. $\mathbb{R}\cap(0,1)$ *and $\mathbb{R}$ have the same cardinality.*

Proof. Consider the function $f : (\mathbb{R}\cap(0,1)) \to \mathbb{R}$ given by $f(x) = \tan(\pi(x+1/2))$. It is not difficult to verify that f is a bijection: it is monotonically increasing, it does not jump ('continuity' being the technical term) and it goes to $-\infty$ and ∞. $\quad\square$

Fig. 11.8: The set $(-\infty,1)\cup(1,\infty)$ has the same cardinality as the semicircle. By drawing a line from a point in $(-\infty,1)\cup(1,\infty)$ to the 'north pole' N, we find exactly one point on the semicircle, the bijection sends A to B and D to C.

There is also a geometric argument (the 'stereographic projection') that constructs a one-to-one mapping between the all real numbers outside of the unit interval $(-\infty, 1) \cup (1, \infty)$ and the semicircle. We see that infinite sets that are contained in a bounded region of space may well contain the same number of elements as infinite sets that are unbounded. It would certainly be interesting to hear what Richard Swineshead or Galileo Galilei would have to say about that construction!

A minor aspect that one might want to worry about is that, *technically speaking*, Cantor's proof is not entirely complete: difficulties could arise from the fact the decimal expansion is not necessarily unique and that, for example

$$0.99999... = 1.00000...$$

The author has been asked about this at some of the more boring parties he attended. People have surprisingly strong opinions about this fact. It has been the experience of the author that the following argument, despite being fairly straightforward, will not usually convince people – it may, in fact, make people angry and can turn a boring party into a slightly less boring party.

Theorem. *We have*
$$0.999999\cdots = 1.$$

Proof. We give two proofs. Let us call $x = 0.99999\ldots$. Then $10x = 9.99999\ldots$. Subtracting one from the other, we get $10x - x = 9$ implying that $x = 1$. That is the end of the first proof. The other proof writes

$$0.999\cdots = 9 \cdot 0.1111\cdots = 9 \cdot \left(\frac{1}{10} + \frac{1}{100} + \frac{1}{1000} + \cdots \right).$$

That is simply a geometric series with $q = 1/10$ and appealing to the formula for the geometric series, we get that this is $9 \cdot 1/9 = 1$. $\qquad\square$

The way this little fact could ruin Cantor's argument is as follows. Suppose

$$f(1) = 1.0000000\ldots$$

and all the other $f(2), f(3), \ldots$ never contain the digit 9 on the diagonal. Then we could always choose the digit 9 (since that is different from whatever I see on the diagonal) and end up with $x = 0.9999\cdots = 1 = f(1)$, which would not give us a contradiction. However, this is a minor issue and can be easily avoided (avoid 0's and 9's). Cantor's 1891 article is rather short, barely 4 pages, and finishes with a beautiful sentence

The further investigation of this area is a task for the future. (Cantor [43])

11.7 Countable vs. uncountable

In the previous subsection, we discussed that infinite sets can have the same cardinality (such as the integers $\mathbb{N}$ and the squares $\{1, 4, 9, 16, \dots\}$) but that they can also have different cardinalities (such as $\mathbb{N}$ and $\mathbb{R}$). At this point, it is convenient to introduce two new words.

> **Definition.** If an infinite set A has the same cardinality as $\mathbb{N}$, it is said to be *countable*. Otherwise, it is *uncountable*.

The choice of words is motivated by the following way of looking at things: if A is countable, then there exists a bijective mapping $f : \mathbb{N} \to A$ in which case $f(1), f(2), f(3), \dots$ is going to be an infinite list of elements in A, and each element in A is going to show up somewhere on the list. We can thus quite literally 'count' A: $f(1)$ being its first element, $f(2)$ its second, and so on. $\mathbb{N}$ is countable. The set of squares $\{1, 4, 9, 16, \dots\}$ is countable. The set of real numbers $\mathbb{R}$ is uncountable. Countable sets have some truly wonderful properties. Suppose we have two infinite sets A and B, and they are both countable: their union $A \cup B$, the collection of all elements that are either in A or in B or in both, is also countable. One way of thinking about a set being countable is that one can enumerate all its elements, the set A can be written as $A = \{a_1, a_2, a_3, a_4, a_5, \dots\}$. If B is also countable, then one can also write it as $B = \{b_1, b_2, b_3, b_4, b_5, \dots\}$. It is perhaps clear how one can assemble their union in a list. $A \cup B = \{a_1, b_1, a_2, b_2, a_3, b_3, a_4, b_4, \dots\}$ is a new list and contains every element that is either in A or in B. This trick extends if we have three countable sets A, B, C, in that case simply arrange the elements as $a_1, b_1, c_1, a_2, b_2, c_2, a_3, \dots$ and so on. Amazingly, the same argument still works if we have *countably many countable sets*!

Theorem. *Suppose $A_1, A_2, A_3, \dots$ is an infinite list of countable sets. Then their union, the set that contains all elements that show up in at least one A_i,*

$$A_1 \cup A_2 \cup A_3 \cup \cdots \qquad \textit{is also countable.}$$

Proof. Each set A_i is countable. We can put its elements into a list. We denote the elements in A_1 by $a_{11}, a_{12}, \dots$, the elements in A_2 by $a_{21}, a_{22}, \dots$ and so on. More generally, a_{ij} will refer to the j-th element in the i-th list. Simultaneously, there are countably many sets $A_1, A_2, \dots$, and we can put those into a list as well. Enumerating both lists, the list of sets and then the elements in each set, we arrive at something like this

$$
\begin{array}{cccccc}
a_{11} & \to & a_{12} & a_{13} & \to & a_{14} & \cdots \\
 & \swarrow & & \nearrow & & \swarrow & \\
a_{21} & & a_{22} & a_{23} & & a_{24} & \cdots \\
\downarrow & \nearrow & & \swarrow & & & \\
a_{31} & & a_{32} & a_{33} & & a_{34} & \cdots \\
 & \swarrow & & & & & \\
a_{41} & & a_{42} & a_{43} & & \cdots & \\
 & & & & & & \\
\cdots & & \cdots & \cdots & & \cdots &
\end{array}
$$

The difficulty lies in combining all these into a single list. We are somehow asked to condense two dimensions (a list of lists) into one dimension (a single list). We tilt our head by 45 degrees and move in a zig-zag fashion through the list: keeping track of elements in the order in which they come along, we arrive at $a_{11}, a_{12}, a_{21}, a_{31}, a_{22}, a_{13}, \ldots$ This new list contains all the elements: in fact, the element a_{ij} shows up within the first $(i + j)^2$ elements of the new list. $\square$

11.8 Applications of Cantor's theory

The last few sections may appear a little bit mystical: infinite sets, cardinalities, some sets are countable and others are not. It may seem as if this was perhaps a bit closer to philosophy than mathematics. Maybe so, but it has some surprising mathematical consequences. We start with a basic observation that will turn out to be very useful.

Theorem. *The set of rational numbers $\mathbb{Q}$ is countable.*

Proof. We already know that $\mathbb{Z}$ is countable. The set of rational numbers with denominator 1, that is

$$A_1 = \left\{ \ldots, -\frac{3}{1}, -\frac{2}{1}, -\frac{1}{1}, \frac{0}{1}, \frac{1}{1}, \frac{2}{1}, \frac{3}{1}, \ldots \right\}$$

is countable (it is really the same as $\mathbb{Z}$). However, the same is true for the set A_2, the set of rational numbers with denominator 2

$$A_2 = \left\{ \ldots, -\frac{3}{2}, -\frac{2}{2}, -\frac{1}{2}, \frac{0}{2}, \frac{1}{2}, \frac{2}{2}, \frac{3}{2}, \ldots \right\}.$$

So A_2 is also countable. The same is true for $A_3, A_4, \ldots$ and where A_k is the list of all rational numbers with denominator k. A countable union of countable sets is countable, and thus $\mathbb{Q}$ is countable. $\square$

There will be some repetition of elements, for example $0 \in A_k$ for all $k \in \mathbb{N}$, but this does not pose a serious issue (we could always delete elements that are repeated). Alternatively, we could relatively directly go through the proof of the Theorem that a countable union of countable sets are countable and adapt the proof

<table>
<tr><td>1/1 1/2 1/3 1/4 ...</td><td></td><td>1/1 1/2 1/3 1/4 ...</td></tr>
<tr><td>2/1 2/2 2/3 2/4 ...</td><td></td><td>2/1   2/3   ...</td></tr>
<tr><td>3/1 3/2 3/3 3/4 ...</td><td></td><td>3/1 3/2   3/4 ...</td></tr>
<tr><td>4/1 4/2 4/3 4/4 ...</td><td></td><td>4/1   4/3   ...</td></tr>
<tr><td>   ⋮ ⋮ ⋮ ⋮ ⋮</td><td></td><td>   ⋮ ⋮ ⋮ ⋮ ⋮</td></tr>
</table>

This gives us a nice way of counting all positive rational numbers $\mathbb{Q}_+$. If they are countable, then $\mathbb{Q}_-$ is also countable and the union of two countable sets is also countable. If someone were to worry about repetition, one could even first remove all repeated rational numbers. One would then end up with the enumeration

$$\mathbb{Q}_+ = \left\{ \frac{1}{1}, \frac{2}{1}, \frac{1}{2}, \frac{1}{3}, \frac{3}{1}, \frac{4}{1}, \frac{3}{2}, \frac{2}{3}, \frac{1}{4}, \ldots \right\}.$$

Despite Pythagoras' hopes, not every number is rational. This was originally shown using $\sqrt{2}$. Using Cantor's ideas, one can give a very different argument: it does not give a concrete example of an irrational number but it shows that they must exist, it actually shows that 'most' real numbers are irrational.

Theorem. *Irrational numbers exist.*

Proof. $\mathbb{Q}$ is countable and $\mathbb{R}$ is not, therefore these two sets have to be different. Since every rational number is real, this means that there exists a real number that is not rational. $\qquad\square$

However, the same type of argument can be used to prove something even stronger. Some numbers are irrational but can still be written nicely as the root of a polynomial with integer coefficients. For example, $\sqrt{2}$ cannot be written as a rational number p/q, but it can be described as

$$\text{the positive solution of} \quad x^2 - 2 = 0.$$

This is a legitimate way of describing real numbers: for example, the real solution of

$$x^5 - x^4 + x^3 - x^2 + x - 2 = 0$$

is given by $x = 1.2148623224884\ldots$ and this real number is uniquely identified by it being the positive, real solution of this equation. We call such numbers *algebraic*.

Definition. We say a real number $x \in \mathbb{R}$ is *algebraic* if there exists a polynomial with integer coefficients $a_n, a_{n-1}, \ldots, a_1, a_0 \in \mathbb{Z}$

$$p(x) = a_n x^n + a_{n-1} x^{n-1} + \cdots + a_1 x + a_0$$

such that $p(x) = 0$. A number that is not algebraic is called *transcendental*.

Every rational number is algebraic, and this is easy to see: if $x = p/q$ with $p, q \in \mathbb{Z}$, then it is a root of the polynomial

$$p(x) = qx - p.$$

Irrational numbers exist (like $\sqrt{2}$). Irrational numbers can also be algebraic (like $\sqrt{2}$). The next natural question is: do transcendental numbers exist? Or is every real number the solution of some polynomial equation with integer equations?

Theorem. *Transcendental numbers exist.*

Proof. The number of polynomials of degree 2 with integer coefficients

$$p(x) = a_1 x + a_2 \quad \text{can be described by} \quad \mathbb{Z} \times \mathbb{Z}$$

since it is enough to keep track of a_1 and a_2. Here, $\mathbb{Z} \times \mathbb{Z}$ refers to the collection of all ordered pairs (a, b) with both $a, b \in \mathbb{Z}$. We observe that $\mathbb{Z} \times \mathbb{Z}$ is countable. Polynomials with three variables

$$p(x) = a_1 x^2 + a_2 x + a_3 \quad \text{can be described by} \quad \mathbb{Z} \times \mathbb{Z} \times \mathbb{Z},$$

where $\mathbb{Z} \times \mathbb{Z} \times \mathbb{Z}$ refers to all triples (a, b, c) with all three elements being integers and so on. So we can describe all polynomials with integer coefficient as a countable union

$$P_1 \cup P_2 \cup P_3 \cup P_4 \cup P_5 \cup \cdots$$

which makes the set of these polynomials countable. Since each polynomial has a finite number of roots, the set of algebraic numbers is countable. Since the real numbers are not countable, *most* real numbers have to be transcendental.   □

11.9 Liouville finds a transcendental number

Cantor's theory shows that transcendental numbers exist and that 'most' real numbers are transcendental. However, proving that any given number is transcendental is actually incredibly difficult. The first explicit construction of a real number x for which transcendence can be proven was carried out by the French mathematician Joseph Liouville (1809–1882). The purpose of this section is to summarize the main idea (without worrying too much about the details). We will then explain why Liouville was also a good friend. His main idea is as follows: *if x is algebraic, then x cannot be extremely well approximated by rational numbers.* More precisely, let us suppose that

$$p(x) = a_n x^n + a_{n-1} x^{n-1} + \cdots + a_1 x + a_0$$

is a polynomial with integer coefficients, and let us assume $p(z) = 0$ for some $z \in \mathbb{R}$. Then this number z is algebraic. Let us now pick a rational number a/b that is very, very close to z but different from z. Then

$$p\left(\frac{a}{b}\right) = a_n \frac{a^n}{b^n} + a_{n-1} \frac{a^{n-1}}{b^{n-1}} + \cdots + a_1 \frac{a}{b} + a_0.$$

A close inspection shows that we can combine the expression on the right-hand side, and, since all variables are integers, we can write

$$p\left(\frac{a}{b}\right) = \frac{\text{some integer } A}{b^n}.$$

It is now true that when a/b is really close to z but different from z, then $p(a/b) \neq 0$. This means that $A \neq 0$ and tells us that

$$\left| p\left(\frac{a}{b}\right) \right| = \frac{|A|}{b^n} \geq \frac{1}{b^n} \qquad \text{is not too small.}$$

The second ingredient that we use is that in a small neighborhood of z, the function $p(x)$ looks a lot like a linear function (this is one of the main ideas behind calculus), and therefore, for some constant $c > 0$, we have

$$|p(x)| = |p(x) - 0| = |p(x) - p(z)| \leq c \cdot |x - z|.$$

Plugging in $x = a/b$ and combining it with the previous inequality, we get

$$\frac{1}{b^n} \leq \left| p\left(\frac{a}{b}\right) \right| \leq c \cdot \left| \frac{a}{b} - z \right|.$$

This tells us that if z is algebraic and the root of a polynomial of degree n. Then, for all rational numbers a/b, we have that a/b cannot be too close to z since

$$\left| \frac{a}{b} - z \right| \geq \frac{1}{c}\frac{1}{b^n}.$$

This inequality is something we have already seen in the case of our friend $z = \sqrt{2}$ (back in Section §7.5). There we showed that

$$\left| \sqrt{2} - \frac{a}{b} \right| \geq \frac{1}{100}\frac{1}{b^2}.$$

This is no coincidence: $\sqrt{2}$ is algebraic because $x^2 - 2$ is a polynomial of degree 2 with integer coefficients for which $\sqrt{2}$ is a solution. Having collected all these ideas, we can now prove the following result.

Theorem (Liouville's constant). *The number*

$$x = \frac{1}{10^{1!}} + \frac{1}{10^{2!}} + \frac{1}{10^{3!}} + \frac{1}{10^{4!}} + \cdots = 0.110001000000000\ldots$$

is transcendental.

Proof. Suppose x is not transcendental. Then it is algebraic with some degree d. The previous considerations show that, for some constant $c > 0$ and all $a, b \in \mathbb{Z}$ with $b \neq 0$

$$\left| x - \frac{a}{b} \right| \geq \frac{1}{c}\frac{1}{b^d}.$$

We will now show that no such inequality can be true. For this, we take the first n terms of the sequence and sum them up: since they are all rational, the sum is rational, and we can set

$$\frac{a}{b} = \frac{1}{10^{1!}} + \frac{1}{10^{2!}} + \frac{1}{10^{3!}} + \cdots + \frac{1}{10^{n!}}.$$

Then $b \leq 10^{n!}$ and the error satisfies, for $n \geq 4$, $|x - a/b| \leq 2 \cdot 10^{-(n+1)!}$. Combining all the inequalities, we get

$$\frac{1}{c}\frac{1}{b^d} \leq \left| x - \frac{a}{b} \right| \leq \frac{2}{10^{n! \cdot (n+1)}} \leq \frac{2}{b^{n+1}}.$$

However, the right-hand side decays only like $1/b^d$ (with d fixed, the degree of the polynomial) whereas the right-hand side decays like $1/b^{n+1}$ which is eventually much smaller than $1/b^d$. This is a contradiction, x has to be transcendental. $\qquad\square$

Liouville's idea is quite general and works for very general infinite sums whose terms decay extremely rapidly. So, for example, the idea could be used to prove that

$$x = 1 + \frac{1}{2^2} + \frac{1}{3^{(3^3)}} + \frac{1}{4^{(4^{(4^4)})}} + \ldots \qquad \text{is irrational.}$$

Liouville was a good friend. He and Jacques Sturm (1803–1855) are now forever linked by *Sturm–Liouville theory* – and they were also friends! They were both in Paris where Sturm, being both a foreigner (originally from Switzerland) and a protestant, was struggling to make a name for himself. One of the hallmarks of scientific importance in Paris is the election to the Academy of Sciences. In 1833, elections were held and the final vote count was as follows

Libri-Carucci : 37   Duhamel : 16   Liouville : 1   Sturm : 0.

Libri-Carucci got elected. Three years later, in 1836, new elections were held. Three weeks before the election, Liouville gave a scientific lecture in front of the Academy in which he *praised* Sturm (as opposed to praising himself).

> Supporting a rival in this way was rather unusual in the competitive Parisian academic circles, and it must have been shocking when on the day of the election, December 5th, Liouville and Duhamel withdrew their candidacies to secure the seat for their friend. Sturm was elected with an overwhelming majority. (Lützen [182])

One final question remains: who is this master mathematician Libri-Carucci who won the 1833 election with such an impressive majority? Guglielmo Libri Carucci dalla Sommaja (1803–1869) was an Italian count and unusual character. After being appointed Chief Inspector of the French Libraries in 1841, he made good use of this position *to steal tens of thousands of books and manuscripts*. In 1848, France issued a warrant for his arrest, but he was tipped off, collected thousands of books and manuscripts, and fled to London where he sustained himself for the rest of his life by selling the books [248].

11.10 Infinitely Many Infinities

We conclude our excursion into the ideas of Georg Cantor with one last spectacular result. We have seen above that the integers $\mathbb{N}$ are contained in the reals $\mathbb{R}$, but that no bijective mapping $f : \mathbb{N} \to \mathbb{R}$ exist. Building on that, we said that $\mathbb{N}$ was countable (the smallest form of infinity) and that $\mathbb{R}$ was uncountable (infinite but larger than $\mathbb{N}$). So maybe there are two types of infinity? The one that is like $\mathbb{N}$ or $\mathbb{Q}$, the countable infinities, and everything else, the uncountable infinities. This is not the case: there are, somehow appropriately, infinitely many different types of infinities. The argument is completely constructive and based on the following insight: for any set, like $\mathbb{N} = \{1, 2, 3, \dots\}$, we can consider subsets: these are sets that contain some of the elements from the original set but not necessarily all of them. For example, if our set is $\mathbb{N}$, then the set of squares $\{1, 4, 9, \dots\}$, the set of even numbers $\{2, 4, 6, \dots\}$ and the set of numbers less than 5, that is $\{1, 2, 3, 4\}$, are all subsets. If we are given a set S, then the set of all subsets, denoted by $P(S)$, is at least as large as S. For any element $s \in S$, the set $\{s\}$ is an element in the power set $\{s\} \in P(S)$. This shows that the power set $P(S)$ cannot be smaller than S. We will show that it is substantially larger: any function $f : S \to P(S)$ has to always map to a strict subset of $P(S)$, it cannot hit all the possible subsets – no bijection exists.

Theorem (Cantor). *If S is a set, then the power set, the set of all subsets, $P(S) = \{T : T \subseteq S\}$, is larger than the set S in the following sense: for $f : S \to P(S)$ there exists a subset $A \subset S$ that is never attained: for all $s \in S$, one has $f(s) \neq A$.*

Proof. The proof has some similarity with the Barber paradox, see §5.3, and it may be a good idea to review it. Let f be any function from S to $P(S)$. We aim to show that f cannot be surjective. The key idea is to realize that every $s \in S$ is mapped to a subset $f(s) \subset S$. It therefore makes sense to ask whether s is contained in $f(s)$ (does the barber shave himself?) and, in particular, to consider

$$M = \{s \in S : s \notin f(s)\}$$

which one may think of as all the people in the village that do not shave themselves. We will now show that, for all $s \in S$, the subset $f(s)$ can never be M. M is the set of all elements $s \in S$ with the property that s is not contained in the subset $f(s)$. Suppose now that $f(t) = M$. Then t either is or is not an element in M.

1. *Case 1.* $t \in M$. If t is in M, then by the definition of M, t is not in $f(t)$ but $f(t) = M$ and this is a contradiction.
2. *Case 2.* If t is not in M, then by the definition of M, t is in $f(t) = M$.

We get a contradiction each time and thus $f(t) = M$ has to be false. $\qquad\square$

11.11 Exercises

1. There is a particularly nice bijection from $\mathbb{N} = \{1, 2, 3, \dots\}$ to the set of squares $\{1, 4, 9, 16, \dots\}$ given by $f(n) = n^2$. Can you find another? Maybe one that does not necessarily have a nice formula attached to it?

2. We discussed Galileo Galilei's discussion of equal cardinalities. Galileo was a fantastic scientist well worth learning about. One of the gems attributed to him is

$$\frac{1}{3} = \frac{1+3}{5+7} = \frac{1+3+5}{7+9+11} = \frac{1+3+5+7}{9+11+13+15} = \cdots$$

Prove this wonderful fact. Nelson [210], who gives a visual proof, refers to the *Galileo Studies* of Drake [72] quoting *The spaces [covered] in accelerated motion from rest and the spaces in uniform motion following accelerated motions, and made in the same times, maintain the same ratio between them* (Galileo [97]).

3. Suppose A, B are two sets and $f : A \to B$ is a bijective function. Show that there exists a function $g : B \to A$ such that, for all $a \in A$, we have $g(f(a)) = a$. Explain why $g : B \to A$ is a bijection.

4. Let A be a finite set containing an odd number of elements. Suppose that $f : A \to A$ has the property that for all $a \in A$, we have $f(f(a)) = a$. Prove that there exists an element $a \in A$ such that $f(a) = a$. Hint: start by looking at an example of a set with 3 or maybe 5 elements.

5. Have a look at the theory that Shakespeare may not have existed (Cantor has a good track record when it comes to being right about things).

6. See whether you can find your own proof that $0.99999\cdots = 1$.

7. Prove that if A is an infinite set, then there exists a function $f : \mathbb{N} \to A$ such that any two distinct integers $m \neq n$ get mapped to different elements in $f(m) \neq f(n)$. Hint: start slowly. Define $f(1)$. See how you would then define $f(2)$.

8. Prove the following special case of Liouville's idea: suppose $x \in \mathbb{Q}$ is a rational number. Then there exists a constant $c > 0$ such that for all rational numbers $a/b \in \mathbb{Q}$ that are different from x, meaning $x \neq a/b$, one has

$$\left| x - \frac{a}{b} \right| \geq \frac{c}{b}.$$

9. In the spirit of Libri-Carucci, is stealing a book always wrong? What if a library is about to throw it out? (The author, to his surprise, was once asked this very question by an undergraduate student working at a Library). What about Abbie Hoffman's book *Steal This Book*?

Chapter 12
The Monte Carlo Method

12.1 Gambling odds

After all the heavy philosophical work in the last section, this section provides some well-deserved relaxation. We discuss a wonderful trick that frequently allows one to avoid cumbersome computation: the Monte Carlo method. It is named after a Casino in Monaco where Stanislaw Ulam's uncle used to gamble. The idea behind the method is well illustrated by the following Stan Ulam story:

> The first thoughts and attempts I made to practice [the Monte Carlo Method] were suggested by a question which occurred to me in 1946 as I was convalescing from an illness and playing solitaires. The question was what are the chances that a Canfield solitaire laid out with 52 cards will come out successfully? After spending a lot of time trying to estimate them by pure combinatorial calculations, I wondered whether a more practical method than "abstract thinking" might not be to lay it out say one hundred times and simply observe and count the number of successful plays. (Ulam [77])

And that's the entire idea! When trying to understand how likely something is, just try it out a couple of times and check. If I want to know the likelihood of me having the winning in hand in poker with a certain set of two cards, I could either try to compute it precisely or I could play 100 games with these two cards. If I have the winning hand 32 out of the 100 games, well, then 32% is not a bad estimate for the probability of winning. This idea is broadly applicable. Suppose, for example, we try to compute the area of the shaded figure in Fig. 12.1.

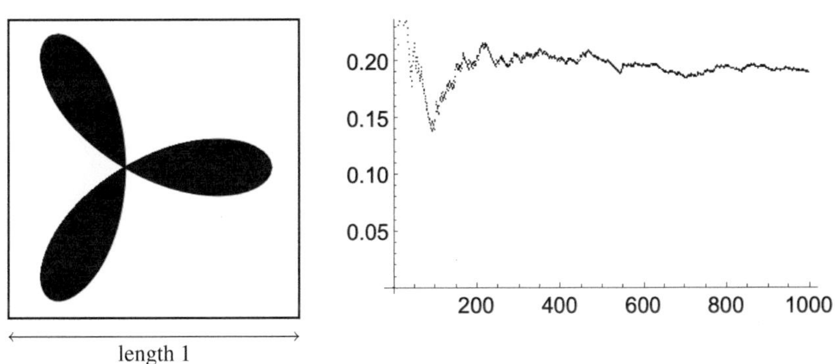

Fig. 12.1: Estimating the area of a shape with Monte Carlo.

© The Author(s), under exclusive license to Springer Nature Switzerland AG 2025   127
S. Steinerberger, *The Unreasonable Elegance of Mathematics*,
Springer Undergraduate Mathematics Series, https://doi.org/10.1007/978-3-032-03815-9_12

One could try to do some sophisticated mathematics, integrals may be involved and it would take a while. On the other hand, there is the Monte Carlo way: if we drop a point randomly in the square, then the likelihood of it hitting the shape will be directly proportional to the area of the shape. If the shape fills out 47% of the square, then a randomly chosen point is inside the domain with a likelihood of 47%. So how about we drop the 1000 points $p_1, \ldots, p_{1000}$ randomly and see how many land in the shape? We define

$$x_i = \begin{cases} 1 & \text{if } p_i \text{ is in the shape} \\ 0 & \text{otherwise.} \end{cases}$$

Obviously, this is not something that we would like to do without a computer. After having dropped the first k points, our best estimate for the likelihood of hitting the shape is

$$\frac{x_1 + x_2 + \cdots + x_k}{k},$$

and that is exactly the figure that is plotted for $k = 1, 2, \ldots, 1000$ in Fig. 12.1. We see that it stabilizes somewhere around 20%, which then suggests that the area of the shape is approximately 0.2.

12.2 Estimating the error: in practice

The Monte Carlo method is randomized: it uses random points, and the output itself is random. If we repeat the same experiment twice, we are going to get two different answers – and maybe the answer is actually far away from the real result? For example, when estimating the area of the shape, the estimate we would have gotten after dropping 100 random points would have been 0.15 (see the dip on the right of Fig. 12.2). There is an amazingly simple solution

If you want to know the accuracy of a Monte Carlo estimate, just repeat it 5 times.

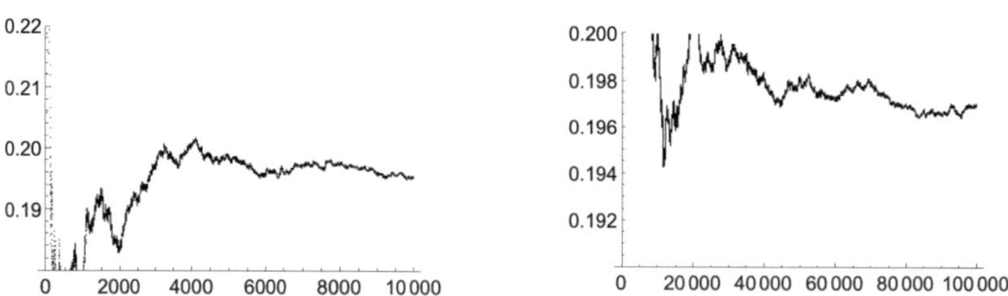

Fig. 12.2: Estimating areas with Monte Carlo, 10,000 and 100,000 points.

We illustrate this for the area problem: dropping 1000 points five times, the number of random points that end up in the shape comes out to be $199, 223, 210, 194$ and 187. This suggests that the area is somewhere between 0.187 and 0.223. It also tells us the approximate accuracy: something like 0.04. Using more points produces a more accurate estimate. Repeating the experiment with 10,000 points five times, we get that the number of points that end up inside the shape is 1959, 1963, 1995, 2032 and 2035, respectively. This gives us a slightly more accurate idea, the area is probably a little bit larger than 0.1959 and a little bit smaller than 0.2035. Repeating the Monte Carlo experiment with 100,000 points five times, we get 19485, 19531, 19680, 19716 and 19784. A clear picture starts to emerge: the area is probably somewhere around 0.196. Moreover, such experiments are very easy to run.

12.3 Estimating the error: in theory

Even when just doing examples on a computer, it can be helpful to have some notion of what one might expect – in some cases this is absolutely crucial. We start with a concrete example. Election Weekend: Candidate Borman vs. Candidate Poole. We are trying to understand who is more likely to win the election and do this by randomly asking a number of people (just as is done in practice). How many people do we need to ask? In a subtle way, the answer will depend on the final result. If we ask 100 random people on the street and 92 are going to vote for Borman, it seems that Borman is very likely to win. If 80 people say they are going to vote for Poole, it appears that Poole is probably pretty likely to win. However, suppose now that 52 people say they are going to vote for Borman and 48 are going to vote for Poole – what then? It appears that Borman has a slight advantage but it also feels like this might be a tight race.

> *Useful Heuristic.* If the proportion of people who are going to vote for Poole is $0 < p < 1$ and we ask n people, we expect that $p \cdot n \pm \sqrt{n}$ of them are going to vote for Poole.

Put in practice, if Poole is going to get a proportion of $0 < p < 1$ of all the votes on Sunday, and we ask n people randomly, then

$$\frac{\text{number of People who say they vote for Poole}}{\text{total number of people that we asked}} = p \pm \frac{1}{\sqrt{n}}.$$

In the example above, if we ask $n = 100$ people, then $1/\sqrt{n} \sim 0.1$ and we can expect to get a correct result up to $\pm 10\%$. If 72 people claim they will vote for Poole, we expect that Poole will get about 70% of the votes, and the 10% error is not large enough to have us worried. Conversely, if 52 people say they favor Borman, well, that is well within the margin of error. A special case of this principle shows up when tossing coins: suppose we have a fair coin that is equally likely to land on each side. We toss it n times and ask: how likely are we to see k heads and $n - k$ tails? There is a nice answer involving nothing but the Pascal triangle (more precisely, its entries, the binomial coefficients).

Theorem. *If we toss a fair coin n times, then*

$$\text{probability of } k \text{ heads} = \binom{n}{k} \frac{1}{2^n}.$$

Proof. If we toss a fair coin, the outcome is either H (heads) or T (tails). Each outcome is equally likely. A coin has no memory, so each sequence of outcomes is equally likely. HHH and TTT have the exact same probability, same as THT and HTT and TTH. If we toss a coin n times, there are $2 \cdot 2 \cdots 2 = 2^n$ possible outcomes. The remaining question is therefore to count the number of events that have k heads. Each such event can be described as a list of k numbers (the number of the coin tosses that result in heads) out of n numbers. That number is $\binom{n}{k}$. $\square$

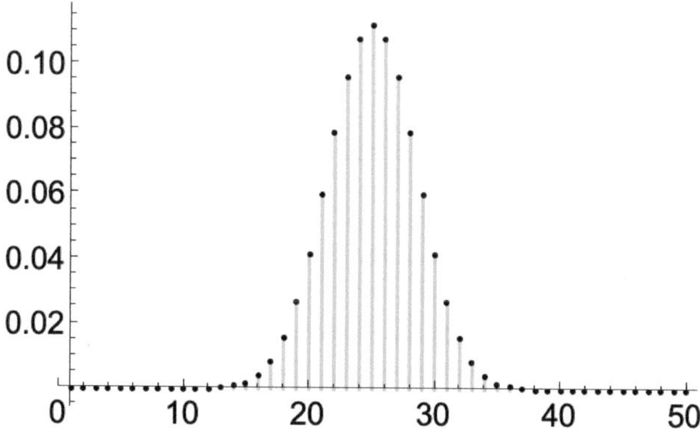

Fig. 12.3: The probabilities of getting k tails when tossing a fair coin 50 times. $k = 25$ is the most likely (approximately 10%), but all numbers between 20 and 30 can be seen somewhat regularly.

If the guiding principle is correct, then that probability should be reasonably large when $k = n/2 \pm \sqrt{n}$ and small otherwise. This is true, but it is not so easy to see this from the formula itself. We simplify using Stirling's formula (see §2.3)

$$n! \sim \sqrt{2\pi n} \left(\frac{n}{e}\right)^n .$$

It is only an approximation but good enough for our purpose. We also make the substitution $k = n/2 + \ell\sqrt{n}$ where $\ell \in \mathbb{R}$ is a parameter. Our goal is to show that the likelihood is reasonably large when $-1 \leq \ell \leq 1$ and that it becomes very small when $|\ell|$ becomes very large. Abbreviating

$$\mathbb{P} = \text{probability } [n/2 + \ell\sqrt{n} \text{ heads}]$$

and using the previous Theorem and the Stirling approximation (three times)

$$\mathbb{P} = \binom{n}{n/2 + \ell\sqrt{n}} \frac{1}{2^n} = \frac{n!}{(n/2 + \ell\sqrt{n})!(n/2 - \ell\sqrt{n})!} \frac{1}{2^n}$$

$$= \text{some computations} \sim \frac{\sqrt{2}}{\sqrt{\pi n}} e^{-2\ell^2} .$$

This formula is very nice because $e^{-2\ell^2}$ is essentially ~ 1 when $|\ell| \leq 1$ and then becomes very small when ℓ becomes large. This confirms exactly what we have been claiming above: when a phenomenon happens 50% of the time (say, seeing heads) and we observe n random instances (coin tosses), we expect to see the thing we are interested in (heads) roughly $n/2 \pm \sqrt{n}$ times. The argument actually gave us a more precise result involving an exponential function. This is the famous Gaussian distribution that can be found everywhere in probability theory: we note again, as we have before for the factorials $n!$ and the binomial coefficients $\binom{n}{k}$, there is a π in the formula. Why does π govern the fluctuations of random coin tosses?

> There is a story about two friends, who were classmates in high school, talking about their jobs. One of them became a statistician and was working on population trends. He showed a reprint to his former classmate. The reprint started, as usual, with the Gaussian distribution and the statistician explained to his former classmate the meaning of the symbols for the actual population, for the average population, and so on. His classmate was a bit incredulous and was not quite sure whether the statistician was pulling his leg. "How can you know that?" was his query. "And what is this symbol here?" "Oh," said the statistician, "this is pi." "What is that?" "The ratio of the circumference of the circle to its diameter." "Well, now you are pushing your joke too far," said the classmate, "surely the population has nothing to do with the circumference of the circle." (Eugene Wigner, The Unreasonable Effectiveness of Mathematics in the Natural Sciences [299])

12.4 Parerga and paralipomena

The replication crisis. The replication crisis refers to an ongoing issue in the sciences where certain experiments cannot be replicated. This is partially to be expected, you do an experiment, and then someone else does it better, and you realize that there was a problem in your original setup and so on – and ultimately we all learn from it. The use of the word *crisis* stems from the fact that *many* experiments, especially in psychology and medicine, are affected. There are many reasons: statistics is difficult, nature is complicated and the incentive structure in academia is convoluted. Reporting an incorrect but spectacular result may lead to immediate career advantages. Another issue is **sample size**. Consider 'An apple a day keeps the doctor away'. This is easy to test, feed apples to a number of people and see whether they need medical attention. At some point, someone will require a doctor and we have disproved the statement. However, most statements are not like that, they are rarely ever all-or-nothing, they are usually more gradual. Suppose I claim that 67% of people have a (resting) heart rate of 65 beats per minute or less. Suppose I find 100 willing participants, 62 of them have a resting heart rate of 65 or less. Does that confirm or disprove the hypothesis? Having n samples allow us to estimate percentages up to an error of $\pm 1/\sqrt{n}$. Having $n = 100$ participants gives an accuracy of $\sim 10\%$ which is not *very* accurate. Analyzing $n = 1000$ participants, giving an accuracy up to $\sim 3\%$, would take *ten times* as long. The author has seen psychological experiments with $n = 12$. Sometimes there are good reasons for this. Think of some extremely rare phenomena, then finding $n = 12$ suitable participants may be a lot of work. However, in general, *beware of small sample sizes*!

A historical footnote. The Monte Carlo method is usually said to have been invented by Stanislaw Ulam in the 1940s – but maybe he was not the first. Enrico Fermi (1901–1954), the creator of the world's first artificial nuclear reactor and a Physics Nobel prize winner, was known for his ability to do very accurate predictions very quickly. *As a person, Fermi seemed simplicity itself. He was extraordinarily vigorous and loved games and sport. [...] This leadership and self-assurance gave Fermi the name of "The Pope" whose pronouncements were infallible in physics. He once said: "I can calculate anything in physics within a factor 2 on a few sheets"* [37]. His secret may have been the Monte Carlo method!

> Fermi took great delight in astonishing his Roman colleagues with his remarkably accurate, "too-good-to-believe" predictions of experimental results. After indulging himself, he revealed that his "guesses" were really derived from the statistical sampling techniques that he used to calculate with whenever insomnia struck in the wee morning hours! And so it was that nearly fifteen years earlier, Fermi had independently developed the Monte Carlo method. (Metropolis [195])

Another historical footnote. There is another historical footnote that illustrates, simultaneously, the Monte Carlo method (in 1901, even earlier than Fermi!) and the replication crisis. It is not too much of a surprise that we can use the Monte Carlo method to approximate π. Figure 12.4 (left) shows a quarter disk inside a unit square: the quarter disk has area $\pi/4$. This means that if we drop n points randomly in the unit square and if m end up inside the quarter disk, then we expect $\pi/4 \sim m/n$ or

$$\pi \sim \frac{4m}{n}.$$

Trying this out on a computer with $n = 1000$ random points, we get $m = 793$ and $4m/n = 3.172$ which is a reasonably good approximation for $\pi = 3.1415\ldots$.

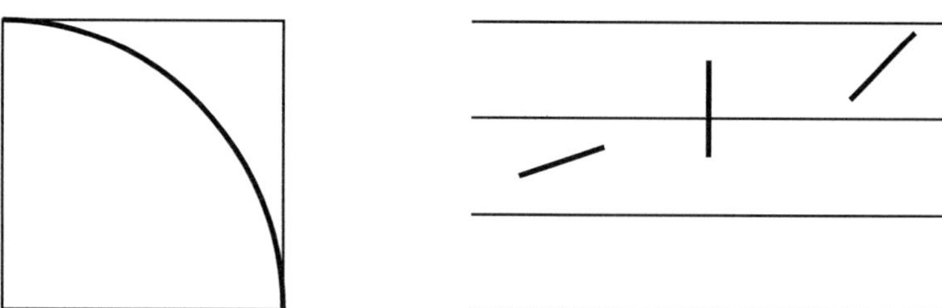

Fig. 12.4: Two ways of computing π via Monte Carlo. Left: $\pi/4$ is the area of a quarter disk, right: dropping little lines of length 1 on a lined piece of paper.

Another method was suggested by Comte de Buffon (1707–1788) who asked a simple question: if we have ruled paper (with the lines being distance 1 apart) and if we drop matchsticks of length 1 randomly on the paper, what is the likelihood of intersecting one of the lines? The answer is $2/\pi \sim 0.63\ldots$, so it happens in roughly 63% of all cases. There is a short argument for people who like integrals. If the line segment has angle α with the x-axis, then its height is $\sin(\alpha)$ and the likelihood of intersecting another line is $\sin(\alpha)$, and thus

$$\text{the likelihood is} \qquad \frac{1}{\pi/2} \int_0^{\pi/2} \sin(\alpha)d\alpha = \frac{2}{\pi} \sim 0.63\ldots$$

We can also turn this into a Monte Carlo experiment: if we drop n line segments of length 1, we would expect that $\sim (2/\pi)n$ of them cross a line and therefore, if we drop n line segments and m of them cross a line, then $\pi \sim 2n/m$. Mario Lazzarini did exactly that in 1901. Lazzarini reports that he got really, really lucky: after dropping 3408 line segments, exactly 1808 of those ended up crossing a line (the numbers do not quite match with our setup since he used line segments whose length is different from the gap size). The ratio he ends up with is

$$\pi \sim \frac{355}{113} = \mathbf{3.1415929}203\ldots$$

which is known to be an *incredibly* accurate approximation to π, the error is $\sim 2 \cdot 10^{-7}$. Do we really believe that Lazzarini was this lucky? This leads straight into the replication crisis! People are doubtful, *something seems a little suspect about those numbers 3408 and 1808. Why cast 3408 needles? Why not a nice round number like 1000 or 3500?* [19] Another funny objection [306] is that Lazzarini reports length and gap as 2.5cm and 3cm, respectively. However, around 1900 the most accurate measurements had an error of $\pm\, 0.0005$cm. Incorporating these bounds, we only get $3.1404 < \pi < 3.1427$, see [19]. Lucky or fraud?

A true story. Many years ago, I met a math PhD who was working as a high-powered financial analyst for a well-known financial institution. I asked him how he used math in his job and got a funny reply: *Every week they come to me with a very tough problem. I start a Monte Carlo simulation on Monday and then spend the week watching TV. On Friday I give them the result and because of the $\pm 1/\sqrt{n}$ law of Monte Carlo my answer is always within a two percent margin of error!* He was having the time of his life. Two years later I ran into him again and asked whether his job was still great and he was completely miserable: *They liked my work so much that they promoted me and gave me one of these fancy offices with glass walls. Now everyone can see what I am doing all the time and I have to actually work!*

12.5 Exercises

1. Maybe you can predict the future. It is not very likely but who knows? Science is about pushing the boundaries. Find a coin, try to predict the future, and then toss the coin. Repeat the experiment 10 times, 20 times or 100 times. What sort of an outcome would be indicative of you being able to predict the future? Note that

if you do 100 coin tosses and are wrong 90 times, that would also be impressive (anti-predicting the future or predicting the anti-future).

2. Test some friends, maybe one of them can predict the future (or, alternatively, by hiding the outcome of the coin toss from them, you can test for telepathic powers). Bonus points if you can convince them that you are engaged in serious research.
3. If you toss a coin 10 times, what is the likelihood of getting 6 heads and 4 tails?
4. Approximate the area of the points in the Euclidean plane that satisfy $x^4 + y^2 \leq 1$.
5. A friend asks for advice: they want to do an experiment but only have $n = 29$ participants. How should they proceed? What does it depend on?
6. Repeat the experiment of Buffon/Lazzarini: draw some lines on a piece of paper and drop some matchsticks. How good is the approximation of π that you get?

Chapter 13
Chaos Theory

I saw one who heard the rustling of the heavens. The second could see Plato's Ideas. A third could number Democritus's atoms. (The Chymical Wedding of Christian Rosencreutz, 1616)

13.1 Laplace's Demon

This chapter is about predictability and unpredictability. It will benefit from a discussion about how people tend to think about the universe: governed by physical laws and evolving like clockwork. If we drop a ball from a great height, then the laws of physics can be used to compute in advance how the ball will fall and, assuming nothing else interferes, this is exactly what will happen. This is more or less how we tend to think of the world. When I board an airplane, I am willing to bet my life that it will move according to the laws of physics. However, this leads to a curious problem that was formulated by a number of people, including Pierre-Simon de Laplace (1749–1827). Laplace, one of the most influential mathematicians of all time, also tried a career in politics and ended up being appointed Minister of the Interior of France – he lasted six weeks, leaving Napoleon quite disappointed:

> As for Minister of the Interior, Minister Quinette was replaced by Laplace, a geometer of the first order. He soon showed himself to not even be a mediocre administrator. As soon as he began his work, the mistake was realized. Laplace never saw any question from its proper point of view, he always looked for subtleties and all his ideas were problematic. Finally, he brought the spirit of the 'infinitely small' into administration. (Napoleon [207])

Despite this unimpressive administrative period, Napoleon thought highly of him as a mathematician. Several years later he wrote to Laplace *expressing the satisfaction which I feel whenever I see you producing new works which render more perfect and advance further the first of the sciences, thus contributing to the lustre of our nation* [227]. Laplace's Essay on Probabilities contains one such interesting advance, a simple observation. If a (hypothetical) very smart being knew everything about the current state of the universe, then this entity would also be able to compute all future states by simply applying the natural laws – everything would be predetermined!

> We may regard the present state of the universe as the effect of its past and the cause of its future. An intellect which at a certain moment would know all forces that set nature in motion, and all positions of all items of which nature is composed, if this intellect were also vast enough to submit these data to analysis, it would embrace in a single formula the movements of the greatest bodies of the universe and those of the tiniest atom; for such an

S. Steinerberger, *The Unreasonable Elegance of Mathematics*,
Springer Undergraduate Mathematics Series, https://doi.org/10.1007/978-3-032-03815-9_13

intellect nothing would be uncertain and the future just like the past could be present before its eyes. (Pierre Simon Laplace, *A Philosophical Essay on Probabilities* [163], 1814)

This hypothetical 'intelligence' has since been known as *Laplace's Demon*. Laplace did not mean to suggest that the demon exists – indeed, we are already in trouble if the Demon *could* hypothetically exist. This would pose a serious challenge to the way we think about physical laws and free will: if I believe in physical laws that govern everything absolutely, then my thoughts two minutes from now will be directly determined by the current state of the universe right now. Then, however, my life is pre-determined. So what do we give up? Our belief in a mechanistic, deterministic universe or our belief in free will? A demonic dilemma indeed!

Fig. 13.1: Laplace's demon calculating our fate.

The idea underlying Laplace's demon was known before Laplace [275]. One fascinating character is Paul Heinrich Dietrich, the Baron of Holbach (1723–1789). Starting in 1750, Holbach regularly hosted philosophical discussions that were attended by many important 18th century intellectuals.

No topics were barred there; "that was the place" said Morellet, "to hear the freest, most animated, and most instructive conversation that ever was ... in regard to philosophy, religion, and government; light pleasantries had no place there." [...] when Hume doubted the actual existence of atheists, the Baron assured him, "Here you are at table with seventeen." ([75])

Holbach argued, in his 1770 book *Systéme de la nature ou des lois du monde physique et du monde moral*, that in a whirlwind of dust, each individual molecule is at its place for a reason and, moreover,

> A geometrician, who would exactly know the different forces which act in these two cases, and the properties of the molecules that are moved, could demonstrate, that, according to given causes, each molecule acts precisely as it ought to act, and could not have acted otherwise than it does. (Baron d'Holbach, 1770, see [275])

This is Laplace's demon under a less spectacular name, *a geometrician*! If each molecule is in a certain place at a certain point in time for a concrete reason, well, then a sufficiently intelligent entity, Laplace's demon or Holbach's geometrician, could surely predict everything. These ideas predate Holbach. A similar idea already occurs in the work of Leibniz as early as 1702

> There is no doubt that a man could make a machine which was capable of walking around a town for a time, and of turning precisely at the corners of certain streets. And an incomparably more perfect, although still limited, mind could foresee and avoid an incomparably greater number of obstacles. And this being so, if this world were, as some think it is, only a combination of a finite number of atoms which interact in accordance with mechanical laws, it is certain that a finite mind could be sufficiently exalted as to understand and predict with certainty everything that will happen in a given period. (Leibniz [168])

An important aspect in these observations is not only that everything is predetermined by natural laws but that an intelligent being could also compute it. The idea of all things following a certain path can be traced back all the way to the Greek philosopher Democritus (460BC–370BC). Democritus is credited with the idea of atoms. Indeed, many of his theories are remarkably accurate.

> Now his principal doctrines were these. That atoms and the vacuum were the beginning of the universe; and that everything else existed only in opinion. That the worlds were infinite, created, and perishable. But that nothing was created out of nothing, and that nothing was destroyed so as to become nothing. That the atoms were infinite both in magnitude and number, and were borne about through the universe in endless revolutions. And that thus they produced all the combinations that exist; fire, water, air, and earth; for that all these things are only combinations of certain atoms; which combinations are incapable of being affected by external circumstances, and are unchangeable by reason of their solidity. Also, that the sun and the moon are formed by such revolutions and round bodies [...] and that everything that happens, happens of necessity. Motion, being the cause of the production of everything, which he calls necessity. (Laertius, [160])

John Tzetzes [285] writes that *As truly being a philosopher who knew all things well, Democritus always used to laugh at the uselessness of life. Now the people of Abdera opined that Democritus was melancholic* and they sent for help! Enlisting none other than *the* physician, Hippocrates, the Father of Medicine after whom the Hippocratic Oath is named, Democritus was healed of his depression. *This Democtritus, then, having been an entirely wise man, Did countless other wonders, they say.* Democritus promptly went to explore the afterlife where he met Hades, the God of Death, *entertaining him with hot blasts from loaves of wheat bread.* Not everybody was as impressed. *And Aristoxenus, in his Historic Commentaries, says that Plato wished to burn all the writings of Democritus that he was able to collect; but that Amyclas and Cleinias, the Pythagoreans, prevented him, as it would do no good; for that copies of his books were already in many hands* [160].

13.2 The Butterflies of Edward Lorenz

Suppose the laws of nature govern absolutely. If you ask a physicist whether this is true, you will get a complicated answer maybe involving quantum mechanics. We assume a simple classic Newtonian universe: objects fall down, pendulums swing and the Earth moves around the Sun. Having agreed that the laws of nature are absolute, there is a second question: how sensitive are they to small changes? This morning I got up and had a cup of coffee. Suppose an incredibly powerful being, maybe Laplace's demon, could replicate the universe exactly as it was today at 6am and change things so that I ended up drinking two cups of coffee instead of one. This will change some things: for example, I may read the newspaper a bit later. Could it change things so much that at noon, a mere six hours later, the world is very different from the way it would be if I had only had one cup of coffee? Probably not, my coffee drinking habits are not of cosmic significance. However, we do not know for sure – and this question is frequently explored in science fiction (especially if time travel is involved). The general idea is widely known under the name 'Butterfly Effect', named after the title of the 1972 paper *Predictability: Does the Flap of a Butterfly's Wings in Brazil Set Off a Tornado in Texas?* written by Edward Lorenz [177]. When it comes to great titles, this one is a classic. Lorenz immediately clarifies (*'Lest I appear frivolous'*)

> The question which really interests us is whether they can do even this—whether, for example, two particular weather situations differing by as little as the immediate influence of a single butterfly will generally after sufficient time evolve into two situations differing by as much as the presence of a tornado. (Lorenz [177])

The question is not merely theoretical, Lorenz points out, *The connection between this question and our ability to predict weather is evident. Since we do not know exactly how many butterflies there are, nor where they are located, let alone which ones are flapping their wings at any instant, we cannot, if the answer to our question is affirmative, accurately predict the occurrence of tornados at a sufficiently distant future time* [177]. As usual, these ideas are much older (though perhaps phrased in a less concrete manner) – everything is connected; for example, the German philosopher Fichte (1762–1814) writes

> In every moment of its duration is nature connected [...] you would be unable to move a single grain of salt without, perhaps invisible to your eyes, change all parts of the imperceptible whole. (Fichte, *The Vocation of Man* [84])

Another remarkable and much more concrete instance is given by Alan Turing (who we already met discussing philosophy with Ludwig Wittgenstein in 1939). Turing writes in 1950, anticipating Lorenz,

> The system of the 'universe as a whole' is such that quite small errors in the initial conditions can have an overwhelming effect at a later time. The displacement of a single electron by a billionth of a centimetre at one moment might make the difference between a man being killed by an avalanche a year later, or escaping. (Turing, [282])

Chaos Theory is concerned with settings where such a striking difference can truly be observed: situations where moving an electron by a billionth of a centimeter really can measurably cause or prevent an avalanche a year later. Before going into the mathematics of Chaos Theory, we enjoy a quick detour into the main subject of Turing's paper *Computing Machinery and Intelligence*.

13.3 A detour: the Turing test

Turing's 1950 paper *Computing Machinery and Intelligence* contains, among other ideas, the notion of the 'Turing test': a hypothetical scenario to test whether a computer program exhibits signs of artificial intelligence. Turing called it an 'imitation game', a person is chatting with what is either a human or a computer program pretending to be a human; if the human is unable to determine whether they are communicating with a human or a program, the program may be said to be intelligent. One may argue that this test has become problematic with the advent of new large language models that certainly chat in a most convincing manner – however, the Turing test already failed much earlier in the most curious way: in 1966, Joseph Weizenbaum published the computer program ELIZA [297], which identifies key phrases, reformulates them and uses them as a reply. This leads to fascinating conversations such as one provided by Weizenbaum [297]:

> Human: Men are all alike.
> ELIZA: In what way?
> Human: They are always bugging us about something or other.
> ELIZA: Can you think of a specific example?
> Human: Well, my boyfriend made me come here.
> ELIZA: Your boyfriend made you come here.
> Human: He says I'm depressed much of the time.
> ELIZA: I am sorry to hear you are depressed.

When one knows the trick, flipping around sentences and asking questions, it is easy to see what is happening. However, if one is unaware, ELIZA (or DOCTOR as it was also known) may appear to be quite sentient. Apparently it was supposed to emulate a psychotherapist of the Rogerian school. Weizenbaum recalls, *I was startled to see how quickly and how very deeply people conversing with DOCTOR became emotionally involved with the computer and how unequivocally they anthropomorphized it. Once my secretary, who had watched me work on the program for many months and therefore surely knew it to be merely a computer program, started conversing with it. After only a few exchanges with it, she asked me to leave the room [...] clear evidence that the people were conversing with the computer as it if were a person who could be appropriately and usefully addressed in intimate terms* [298].

13.4 Chaos Theory

Time to jump into Chaos Theory. We are interested in a setting where

1. the state of the universe at a fixed point in time determines all subsequent states of the universe for all time, a setting of complete determinism,
2. where the tiniest change can have huge consequences later on.

Our setting will be simple: the state of the universe will be described by a number $0 \leq x \leq 1$. The 'physical laws' are described by a function $f : [0, 1] \to [0, 1]$ that send a number from $[0, 1]$ to another number from $[0, 1]$. We may think of x as the current state of the world today and $f(x)$ as the state of the world tomorrow. This setup is completely deterministic: every time the state of the world is x, it will always be $f(x)$ tomorrow and $f(f(x))$ the day after. We are then interested in how things evolve in time, that is, we will be interested in

$$x, f(x), f(f(x)), f(f(f(x))), f(f(f(f(x)))), \ldots$$

One quickly sees that this is a little bit difficult to read, and we will abbreviate the k-fold application of a function f by $f^{(k)}(x)$. More formally, we define $f^{(1)}(x) = f(x)$ and $f^{(k+1)}(x) = f(f^{(k)}(x))$. There is no need to be overly abstract, we will start with a very concrete function: the doubling map $d(x)$. Its formal definition is

$$d(x) = \begin{cases} 2x & \text{if } 0 \leq x \leq 1/2 \\ 2x - 1 & \text{if } 1/2 < x \leq 1 \end{cases}$$

A graph of the function $d(x)$ is shown in Fig. 13.2. For any fixed initial value

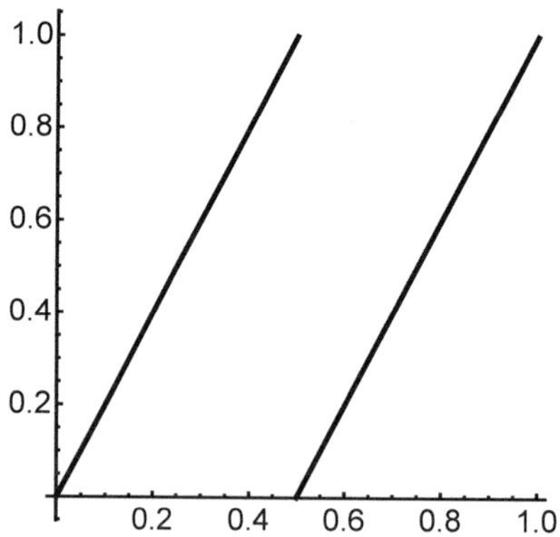

Fig. 13.2: The doubling map $d(x)$

$0 \leq x \leq 1$, the sequence $d(x), d^{(2)}(x), d^{(3)}(x), \ldots$ is well-defined for all times. However, it is also tremendously unstable: if we change x just a little bit, the sequence may be completely different.

Theorem (The doubling map is unstable). *For any $0 < x < 1$ and any $\varepsilon > 0$ there exists $0 \leq y \leq 1$ and $n \in \mathbb{N}$ so that*

1. *the number y is extremely close to the number x*

$$|x - y| \leq \varepsilon$$

2. *but after n steps, the numbers $d^{(n)}(x)$ and $d^{(n)}(y)$ are very different*

$$\left| d^{(n)}(x) - d^{(n)}(y) \right| \geq \frac{1}{4}.$$

One way of thinking about the result is as follows: if we know x exactly, then the future is determined exactly (as we desire: complete determinism). However, if we only know x up to some accuracy, say the first 1000 digits of x (which is much better than what one gets in actual experimental physics), then we can only predict the future for a short period of time: there exists an n such that $d^{(n)}(x)$ is very far away from $d^{(n)}(y)$, where y is a value that is so close to x that we cannot determine whether the universe is currently in state x or state y. In terms of Laplace's demon, it would imply that the demon would have to know the current state of the universe to an infinite degree of accuracy – but perhaps this is not a problem for the demon.

Proof of Instability of the doubling map. The trick is to write the number x in its binary expansion: we write

$$x = \frac{a_1}{2} + \frac{a_2}{4} + \frac{a_3}{8} + \frac{a_4}{16} + \cdots$$

where each a_i is either 0 or 1. This is also sometimes abbreviated as $x = (a_1 a_2 a_3 \cdots)_2$. For example, the number $1/7$ would then be written as

$$\frac{1}{7} = \frac{0}{2} + \frac{0}{4} + \frac{1}{8} + \frac{0}{16} + \frac{0}{32} + \frac{1}{64} + \cdots = (001001001\ldots)_2$$

and we see that this periodic pattern continues: the geometric series tells us that

$$\frac{1}{8} + \frac{1}{64} + \frac{1}{512} + \cdots = \frac{1}{7}.$$

Not all numbers have periodic expansions, but that is not a problem. Let us now suppose that

$$x = \frac{a_1}{2} + \frac{a_2}{4} + \frac{a_3}{8} + \frac{a_4}{16} + \cdots$$

and suppose we want to compute the doubling map $d(x)$. We first need to determine whether $x \leq 1/2$ or $x > 1/2$, and this is actually surprisingly easy: we have $x < 1/2$ if and only if $a_1 = 0$ and $x \geq 1/2$ if and only if $a_1 = 1$.

Case 1. $x < 1/2$ or, equivalently, $a_1 = 0$. Then $d(x) = 2x$ and

$$d(x) = 2x = 2\left(\frac{a_1}{2} + \frac{a_2}{4} + \frac{a_3}{8} + \frac{a_4}{16} + \cdots\right)$$
$$= 2\left(\frac{a_2}{4} + \frac{a_3}{8} + \frac{a_4}{16} + \cdots\right) = \frac{a_2}{2} + \frac{a_3}{4} + \frac{a_4}{8} + \cdots.$$

Case 2. $x \geq 1/2$ or, equivalently, $a_1 = 1$. Then $d(x) = 2x - 1$ and

$$d(x) = 2x - 1 = 2\left(\frac{a_1}{2} + \frac{a_2}{4} + \frac{a_3}{8} + \frac{a_4}{16} + \cdots\right) - 1$$
$$= 2\left(\frac{1}{2} + \frac{a_2}{4} + \frac{a_3}{8} + \frac{a_4}{16} + \cdots\right) - 1$$
$$= 2\left(\frac{a_2}{4} + \frac{a_3}{8} + \frac{a_4}{16} + \cdots\right) = \frac{a_2}{2} + \frac{a_3}{4} + \frac{a_4}{8} + \cdots.$$

Both cases end up in the same final result. We have just seen that

$$\boxed{x = (a_1 a_2 a_3 \ldots)_2 \implies d(x) = (a_2 a_3 a_4 \ldots)_2.}$$

This absolutely wonderful equation tells us that computing $d(x)$ is really easy when x is given in its binary representation. It also implies that

$$\boxed{x = (a_1 a_2 a_3 \ldots)_2 \implies d^{(n)}(x) = (a_{n+1} a_{n+2} a_{n+3} \ldots)_2.}$$

This is all that we need to complete the argument, which is perhaps best done with an example. Let us suppose, for example, that $x = 1/7$, meaning

$$x = (001001001001001001001 \ldots)_2 = \frac{1}{7}.$$

Changing just a single digit many digits after the decimal point leads to a very small change of the number ('moving an electron by a billionth of a centimeter')

$$y = (00100100100100100\underline{1}011 \ldots)_2 = \frac{1}{7} + \frac{1}{1048576}$$

but after applying the doubling map for a sufficient amount of time, we have

$$d^{(19)}(x) = (01001001 \ldots)_2 = \frac{4}{14}$$

while

$$d^{(19)}(y) = (11001001 \ldots)_2 = \frac{11}{14}.$$

These two numbers are very apart (avalanche vs. no avalanche). $\square$

13.5 Beyond the doubling map

The doubling map is quite satisfying: it gives us a nice concrete example of a way that numbers in $[0, 1]$ can be transformed in a deterministic way into other numbers in $[0, 1]$ so that even very slight changes can lead to a completely different outcome. However, one might argue, the doubling map is a very special type of function. There exists a second trick that is worth seeing once. Suppose we have a function like the doubling map $d(x)$ and understand everything about

$$d(x), d^{(2)}(x), d^{(3)}(x), \ldots$$

Assume now that there exists an (invertible) function $h(x)$ and consider the new function $f(x) = h^{-1}(d(h(x)))$, where $h^{-1}(x)$ is the inverse function to $h(x)$. For example, if h is multiplication by 2, meaning $h(x) = 2x$, then the inverse function is division by 2 and $h^{-1}(x) = x/2$. Or, if we are dealing with positive numbers and $h(x) = \sqrt{x}$, then $h^{-1}(x) = x^2$ since that is the inverse operation to taking a square root. Then the behavior of

$$f(x), f(f(x)), f(f(f(x))), f^{(4)}(x), f^{(5)}(x), \ldots$$

is not that different from the behavior of the doubling map. There is an easy way to see this: the first few iterations, using $f(x) = h^{-1}(d(h(x)))$, are

$$f(x) = h^{-1}(d(h(x)))$$
$$f(f(x)) = h^{-1}(d(h(h^{-1}(d(h(x))))))$$
$$f(f(f(x))) = h^{-1}(d(h(h^{-1}(d(h(h^{-1}(d(h(x)))))))))).$$

Something nice happens: $h(x)$ and $h^{-1}(x)$ are inverse functions, so $h(h^{-1}(x)) = x$. Using the example above, if we first multiply by 2 and then divide by 2, we end up with the number that we started with. Alternatively, if we first take the square root of a positive number and then square that, we end up with the original number. Even better, the order does not matter. Going through the equations above and erasing pair of $h(h^{-1}(x))$, the equation simplifies to

$$f(x) = h^{-1}(d(h(x)))$$
$$f(f(x)) = h^{-1}(d(d(h(x))))$$
$$f(f(f(x))) = h^{-1}(d(d(d(h(x)))))$$

and, more generally,
$$f^{(n)}(x) = h^{-1}(d^{(n)}(h(x))).$$

The evolutions of f and d are really the same in a sense: the technical term is that f and d are *topologically conjugated* to each other. If we understand one, we understand the other. And if one is chaotic, then the other one is also chaotic. There is a famous example that illustrates this idea nicely. This example (the *logistic map*) of a chaotic system is given by

$$f(x) = 4x(1 - x).$$

It looks like a very friendly function but the behavior of $f(x), f(f(x)), f(f(f(x)))$ and so on can quickly become quite unpredictable. It may help to illustrate this with a simple example. We have $f(3/4) = 3/4$, a fixed point, which is nice. This is a stable point that does not change in time. If one starts at $3/4$, one will always stay at $3/4$ for all eternity. What if we start at say, $x = 3/4 + 1/10000$? A possible answer is shown in Fig. 13.3: initially, since we are close to the stable point $3/4$, presumably we stay close to the stable point. However, after a few steps, we see a rather complicated behavior all over the place in $[0, 1]$.

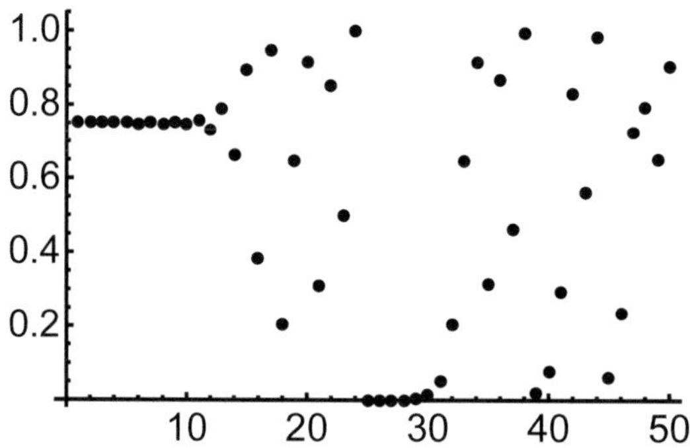

Fig. 13.3: *Maybe* the first 50 steps of $f(x), f^{(2)}(x), \ldots$ when $x = 3/4 + 1/100000$.

Fig. 13.3 illustrates another difficulty with these chaotic systems: in order to accurately predict them for all time, we have to know the initial value precisely. This does not appear to be a problem since we know everything about $x = 3/4 + 1/100000$. However, a computer does not and a computer is certainly no Laplace's Demon. A computer, unless otherwise instructed, will only save a number up to 16 digits after the decimal point. So can we really trust that the computer is doing the correct thing? This is, in fact, how this phenomenon was discovered when Lorenz tried to predict the weather using a computer: he ran the computations and then ran them again.

At one point I decided to repeat some of the computations in order to examine what was happening in greater detail. I stopped the computer, typed in a line of numbers that it had printed out a while earlier, and set it running again. I went down the hall for a cup of coffee and returned after about an hour, during which time the computer had simulated about two months of weather. The numbers being printed were nothing like the old ones. I immediately suspected a weak vacuum tube or some other computer trouble, [...] I found that the new values at first repeated the old ones, but soon afterward differed by one and then several units in the last decimal place, and then began to differ in the next to the last place and then in the place before that. In fact, the differences more or less steadily doubled in size every four days or so, until all resemblance with the original output disappeared somewhere in the second month. This was enough to tell me what had happened: the numbers that I had typed in were not the exact original numbers, but were the rounded-off values that had appeared in the original printout. The initial roundoff errors were the culprits; they were steadily amplifying until they dominated the solution. In today's terminology, there was chaos. (Edward Lorenz, *The Essence of Chaos*, [178])

Lorenz points out that this has direct relevance for weather forecasting since weather behaves in a similar way. *It soon struck me that, if the real atmosphere behaved like the simple model, long-range forecasting would be impossible. The temperatures, winds, and other quantities that enter our estimate of today's weather are certainly not measured accurately to three decimal places, and, even if they could be, the interpolations between observing sites would not have similar accuracy* [178]. This analogy has turned out to be fairly accurate. Weather does seem to behave in a chaotic way and we are not *much* better at predicting it than we were 20 years ago. This is not just a computational issue: even if we covered the entire surface of the Earth with weather sensors that measure everything with an accuracy of 1 in a million, we would still not be able to predict the weather for longer than a few weeks. Time for the main result of this section: the type of chaotic behavior that we saw for the doubling map, small changes in the starting value leading to completely different behavior later on, also occurs for the logistic map $f(x) = 4x(1 - x)$.

Theorem. $f(x) = 4x(1 - x)$ *is topologically conjugate to the doubling map.*

Proof. The doubling map is given by

$$d(x) = \begin{cases} 2x & \text{if } 0 \leq x \leq 1/2 \\ 2x - 1 & \text{if } 1/2 < x \leq 1. \end{cases}$$

Using the trick from the preceding section, we want to find a function $h(x)$ such that $f(x) = h(d(h^{-1}(x)))$. Replacing $x = h(y)$, we are led to the problem of finding an h such that $f(h(y)) = h(d(y))$. Here it turns out, as if by magic, the function $h(x) = \sin(2\pi x)^2$ does the trick. Writing things out and after some computation

$$f(h(y)) = 4\sin(2\pi y)^2(1 - \sin(2\pi y)^2) = \sin(2\pi d(y))^2 = h(d(y)). \qquad \square$$

There is one part of the argument that is perhaps a little bit unsatisfying: where does $h(x) = \sin(2\pi x)^2$ come from? The answer is: hard work and a little bit of luck, finding $h(x)$ is not always easy.

13.6 Chaos in the solar system

Since we are all living in the solar system, it would be nice to understand what it is going to do next. Surely, if we knew the position and speed of all the planets today, then the laws of nature would be able to predict the location (and speed) of the planets tomorrow and the day after and really as long as we wish.

QUATORZIÉME MÉMOIRE.

Réfléxions fur le Problème des trois Corps, avec de nouvelles Tables de la Lune, d'un ufage très-fimple & très-facile.

Fig. 13.4: d'Alembert's Fourteenth Memoire (1761): *Thoughts on the problem of three bodies, with new tables of the moon that are very simple and easy to use* [2].

Things, once more, are not so simple. The physical laws describing the motion of planets were already written down by Newton in the 17th century and are widely agreed upon (Einstein had some small corrections later on but let's ignore that for now). If we pretend that there are only two celestial objects, say the Sun and the Earth, these equations can be nicely solved. However, it was quickly realized that as soon as there are three celestial objects (Sun, Earth and Moon, for example), then the equations do not have a 'nice' solution like $f(x) = x^2$. They still have a solution, but they can only be written down using infinitely many sums and that is complicated to analyze. This has become known as the 'three-body problem' (see Fig. 13.4). A natural way out would be to argue as follows: the Sun is a lot bigger and a lot heavier than all the other planets, maybe the way the Earth moves around the Sun is actually pretty much independent of whether or not Neptune exists: Neptune is much smaller than the Sun and far away. Maybe each planet moves happily around the Sun in a way that they nearly ignore each other. The exact opposite is true – that's how Neptune's existence was first suspected:

> With M. Alexis Bouvard I had some conversation upon a subject I had often meditated, which will probably interest you, and your opinion may determine mine. Having taken great pains last year with some observations of Uranus, I was led to examine closely Bouvard's table of that planet. The apparently inexplicable discrepancies between the ancient and modern observations suggested to me the possibility of some disturbing body beyond Uranus, not taken into account because unknown. (Letter from T. Hussey to G. Airy, 1834 [1])

Indeed, the planet formerly known as Pluto needs 247.94 years to orbit the Sun, Neptune (being closer) requires only 164.8 years. Note that

$$\frac{247.94}{164.8} = 1.50449 \sim \frac{3}{2}.$$

This may be a coincidence. Inspired by Leurechon who noted that *4 stars have been detected that move around Jupiter* [171], we may consider the three moons Ganymede, Europa and Io; their orbital periods around Jupiter are 7.154, 3.551 and 1.769 days, respectively. Note that

$$\frac{7.154}{1.769} \sim 4.04 \qquad \frac{7.154}{3.551} \sim 2.014 \qquad \text{and} \qquad \frac{3.551}{1.769} \sim 2.007.$$

This is *not* a coincidence and part of a larger issue ('orbital resonance'). Thus, when modeling planetary motion, we cannot simply treat Sun–Earth and Sun–Jupiter in isolation, Earth and Jupiter are going to interact as well. So what is going to happen to the solar system in the long run? It's an interesting question, and in 1885 the King of Sweden, Oscar II, decided to turn it into a *competition* (Fig. 13.5).

His Majesty Oscar II., wishing to give a fresh proof of his interest in the advancement of mathematical science, an interest already manifested by his graciously encouraging the publication of the journal *Acta Mathematica*, which is placed under his august protection, has resolved to award a prize, on January 21, 1889, the sixtieth anniversary of his birthday, to an important discovery in the field of higher mathematical analysis. This

Fig. 13.5: Prize Announcement in *Nature* (July 1885) [200]

The competition asks to consider *A system being given of a number whatever of particles attracting one another mutually according to Newton's law, it is proposed, on the assumption that there never takes place an impact of two particles, to expand the coordinates of each particle in a series proceeding according to some known functions of time.* It is also pointed out that solving this problem *will considerably enlarge our knowledge with regard to the system of the universe* [200]. One of the submissions to this concept was by Henri Poincaré (1854–1912). It did not solve the problem (so far, nobody has), but it started an entirely new area of mathematics. Henri Poincaré was an amazing scientist. In addition to doing lots of beautiful mathematics, he also discovered Lorentz transformations three months before Einstein's first paper on special relativity. While perhaps less known today than he deserves, Poincaré was well regarded in his time. Indeed, in 1910 the psychologist Édouard Toulouse wrote an entire book about the way in which he approached problems and everything else: we learn that *He is not a hunter, His sleep is often poor with daily interruptions lasting 2 or 3 hours, he does not have the gift of miming* and *he does not play chess and believes that he would not make a good player* [281] and many other things. Returning to the solar system, the problem of what will happen to it in the next billion years has never been solved and is perhaps unsolvable. Some newer work suggests that it will continue in its present shape for at least several million years, so not an excuse not to file taxes.

13.7 Exercises

1. Contemplate Laplace's demon and what it means for the belief in a deterministic universe vs. the existence of free will. Which belief would you prefer to sacrifice?
2. People like to debate whether they believe in 'free will'. One interesting aspect of such debates is that people often have a very unclear notion of what they mean by 'free will'. See whether you can make this notion precise (there may not be a unique answer here).
3. If you believe in free will (whatever that may be) and want to challenge your belief, have a look at the work of Deecke and Kornhuber about the 'readiness potential' [158]. Scary, no?
4. Repeat Lorenz' experiment: take a nice irrational number, maybe $x = 1/\sqrt{2}$, type it into a calculator and compute $x, d(x), d^{(2)}(x), d^{(3)}(x), \ldots$ and then do the same on a different software system. Do you end up getting different numbers after a while? (This may depend on how the software system stores numbers).
5. Prove the formula $d^{(k)}(x) = \{2^k x\}$, where $\{x\}$ is again the fractional part (the part of the number after the decimal point). Induction might be useful.
6. Prove that for all $0 < a < b < 1$ there exists $a < x < b$ such that $d(x), d^{(2)}(x), \ldots$ is eventually periodic.
7. *Challenging*: Prove that for all $0 < a < b < 1$ there exists $a < x < b$ such that $d(x), d^{(2)}(x), \ldots$ is eventually periodic with a period k satisfying $k \leq 1000 \log(1/(b - a))$. Can you prove it with a number smaller than 1000?
8. If you *truly* enjoy trigonometric functions, prove that

$$4 \sin\left(2\pi y\right)^2 (1 - \sin\left(2\pi y\right)^2) = \sin(2\pi d(y))^2.$$

Chapter 14
Mathematics as a Tool for Dishonesty

14.1 Social Sciences Sorcery

Mathematics is a powerful tool but, like every tool, it can be abused. There are many reasons to study mathematics: it is beautiful, it is fun, it is interesting. Sadly, being able to defend oneself against the abuse of mathematics is also an important reason. This is particularly relevant when mathematics is employed outside the natural sciences. It is clear that problems in economics, psychology, sociology and so on are intrinsically difficult: the current state of the economy or someone's state of mind cannot be reduced to a single number. Simplifying models have to be used, approximations come into play. Some people welcome this as an opportunity to use mathematics incorrectly. Sometimes that abuse is unintentional, everybody makes mistakes, and sometimes it is very intentional. Typical signs of the latter include:

1. Equations are introduced in a casual way. *Obviously*, these are the right equations. It helps that many equations look really, really scientific [149]. There is an element of intimidation, the reader must be a fool for disagreeing [150].
2. Whether the equations actually apply or whether they are merely approximations, is either not discussed at all or discussed in passing.
3. Advanced mathematical theories may be used; these can serve as another insurance against criticism: only someone well versed in these mathematical theories can challenge their application on a formal level.

A less polite but not necessarily inaccurate formulation of the problem was given by the sociologist Stanislav Andreski in 1972.

> The recipe for authorship in this line of business is as simple as it is rewarding: just get hold of a textbook of mathematics, copy the less complicated parts, put in some references to the literature in one or two branches of the social studies without worrying unduly about whether the formulae which you wrote down have any bearing on the real human actions, and give your product a good-sounding title, which suggests that you have found a key to an exact science of collective behaviour. (S. Andreski, *Social Sciences as Sorcery* [5])

Andreski, not a boring writer, is also perfectly clear in where the overwhelming amount of nonsense comes from and asks us to *Remember that publishers want to keep the printing presses busy and do not object to nonsense if it can be sold. As my grandmother used to say, paper is patient.* Moreover, *clear and logical thinking [...] sooner or later undermines the traditional order. Confused thinking, on the other hand, leads nowhere in particular and can be indulged indefinitely without producing any impact upon the world.* [5]. Since we just covered Chaos Theory, we make this concrete by discussing a striking example: the work of Losada. We recall, from the previous chapter, that Edward Lorenz studied weather formation and used a

S. Steinerberger, *The Unreasonable Elegance of Mathematics*,
Springer Undergraduate Mathematics Series, https://doi.org/10.1007/978-3-032-03815-9_14

computer to solve the equations. When he restarted the system, he discovered that the solutions looked different and that rounding errors became dominant, thus leading to chaotic behavior. It will be useful to look at the actual shape of Lorenz equations at least once, they look as follows

$$\frac{dx}{dt} = \sigma(y - x), \quad \frac{dy}{dt} = x(\rho - z) - y, \quad \frac{dz}{dt} = xy - \beta z$$

$x(t), y(t), z(t)$ are functions depending on time while σ, ρ, β are three numbers. dx/dt is an abbreviation for 'rate of change of $x(t)$'. No worries, for what follows, we do not actually need to know what these equations mean, just what they look like.

14.2 The Losada equations

The story starts with the 1999 paper *The complex dynamics of high performance teams* written by Marcial Losada [179]. It is an interesting document, the paper literally starts *For several years I had been the director of the Center for Advanced Research (CFAR), built by EDS in Ann Arbor and Cambridge, near the University of Michigan and MIT campuses.* This will impress some people. I have never been the director of anything, I have never heard of CFAR or EDS, but it sounds very scientific. I have heard of the University of Michigan and MIT. It is amusing that the center is described as being *near* the campus of the University of Michigan and MIT (as opposed to *on*). The article is concerned with the performance of teams. Losada argues that the *connectivity of a team is highly correlated with its performance*, and this seems very natural: teams that do not communicate well are probably not as effective as teams that do. It is furthermore argued that *Connectivity is measured by the strength and number of cross-correlations among time series of the coded speech acts of meeting participants*, which is perhaps a little bit vague but sounds reasonable: if people talk more with each other, they are probably more connected. Cross-correlation of time series refers to

$$(f \star g)(x) = \int_{-\infty}^{\infty} \overline{f(t)} g(t + x) dt$$

and if that is not proper science then what is? Then things get even more interesting: equations are being derived.

> One such interaction is that between inquiry-advocacy and other-self. If I call the first X and the second Y, their interaction should be represented by the product XY, which is a nonlinear term. I also knew from my observations at the lab, that this interaction should be a factor in the rate of change driving emotional space (which I call Z). In addition, I would need a scaling parameter for Z. Consequently, the rate of change of Z should be written as
>
> $$\frac{dZ}{dt} = XY - aZ$$
>
> (The first equation and its explanation, Losada [179])

This equation look pretty serious. I must confess that it is not quite clear to me what inquiry-advocacy and other-self are. They must be something like real numbers or maybe real-valued functions, because they are being multiplied. I do not understand what is happening but I am also not the Director of a Research Institute located *near* the campus of MIT – surely the author understands everything perfectly well. As for the equation itself, well, that is anyone's guess at this point: it is not explained, and I do not understand it. I have spent my entire life not understanding equations, and, at least historically, this was usually my fault. The story continues.

> From my observation at the lab, I also knew that connectivity had a critical incidence on the level of inquiry-advocacy and, consequently, it should interact with X and the product of this interaction should be part of the rate of change of Y, according to the characteristics of the time series observed, where there was a lead-lag relationship between Y and X. I also needed to discount the interaction between X and Z (which would be represented by the nonlinear term XZ and Y with itself, so that the rate of change of Y should b written as
>
> $$\frac{dY}{dt} = cX - XZ - Y$$
>
> where c is the control parameter representing connectivity, as measured by the nexi index, and should be varied according to the nexi number for each term performance category.

(The second equation and its explanation, Losada [179])

The mystery certainly deepens but okay, it is maybe not so easy to put your finger on something specific and say 'Hey, this is wrong'. It takes a little bit of courage to think that passages in a published scientific article are without meaning. Andreski remarks *if you are a famous man in a top position and with a lot of influence, few people will dare to say, or even think, that it is all nonsense, lest they be accused of ignorance and lack of intelligence, and forfeit their chances of obtaining appointments, invitations or grants* [5]. Losada finishes by deriving a third equation which is so obvious to the expert that no real justification is needed

> Finally, and in accordance with the characteristics of the time series generated at the lab, the rate of change of X should be a function Y, discounting the level of X; so that with the inclusion of a scaling parameter, the rate of change of X should be written as
>
> $$\frac{dX}{dt} = b(Y - X),$$
>
> where b is a scaling parameter to be held constant. (Third equation, Losada [179])

The three Losada equations now look as follows

$$\frac{dX}{dt} = b(Y - X), \quad \frac{dY}{dt} = cX - XZ - Y, \quad \frac{dZ}{dt} = XY - aZ,$$

where a, b, c are numerical quantities that have not yet been specified. As one quickly sees, this looks very much like the Lorenz equations. What a coincidence! Losada shares our excitement and writes *I realized that, except for some differences in the arrangement of the terms and the letters chosen to designate the parameters, these were the same set of coupled nonlinear differential equations that Lorenz had chosen for his model and published in one of the most often cited papers in science* [179]. Truly a miracle! Ask yourself, is the serendipitous discovery that two seemingly different things are really the same, is that not often how science progresses?

14.3 Losada aftermath

The story continues. In 2005, a second paper [88] shows that by extending *Fredrickson's (1998) broaden-and-build theory of positive emotions and M. Losada's (1999) nonlinear dynamics model of team performance, the authors predict that a ratio of positive to negative affect at or above 2.9 will characterize individuals in flourishing mental health* [88]. The story was examined by Brown–Sokal–Friedman, who point out that *It is worth stressing that Fredrickson and Losada (2005) did not qualify their assertions about the critical positivity ratios in any way. The values 2.9013 and 11.6346 were presented as being independent of age, gender, ethnicity, educational level, socioeconomic status or any of the other many factors that one might imagine* [41]. Not without a smile, they point out that *the idea that any aspect of human behavior or experience should be universally and reproducibly constant to five significant digits would, if proven, constitute a unique moment in the history of the social sciences* [41]. Fredrickson has since come *to see sufficient reason to question the particular mathematical framework Losada and I adopted* [89]. She also remarks that *Losada, however, chose not to respond* [89]. While this case is certainly a bit extreme, it is worth emphasizing that many more such cases exist – most of them are much better at hiding what they are doing; many are never challenged.

14.4 Exercises

1. The next time someone writes down a mathematical equation and says that it obviously describes what is happening and you do not agree, speak up. It is not always easy, but it is worth it.
2. Read Losada's original paper [179] and mark every sentence that uses technical terms that were never properly defined (*connectivity had a critical incidence on the level of inquiry-advocacy*).

Chapter 15
Able Amateurs and Colorful Crackpots

Mathematics requires very little in terms of actual resources: one does not need a lab, one does not need a particle collider, and, usually, one does not need an entire library. Sometimes the right book can make a difference. A computer can be useful. This leads to two interesting developments (that have analogous forms in other formal sciences) which we will, tongue-in-cheek, refer to as

1. the able amateur and
2. the colorful crackpot.

15.1 Able Amateurs

Benjamin Franklin. Benjamin Franklin (1706–1790) was an American polymath. He did not hold mathematicians in high regard. His Junto club, founded as a club for mutual improvement in 1727, had Thomas Godfrey (1704–1749) as a founding member whom Franklin describes as follows

> Thomas Godfrey, a self-taught mathematician, great in his way, and afterward inventor of what is now called Hadley's Quadrant. But he knew little out of his way, and was not a pleasing companion; as like most great mathematicians I have met with, he expected universal precision in everything said, or was forever denying or distinguishing upon trifles, to the disturbance of all conversation. He soon left us. (Benjamin Franklin [87])

Franklin recalls that in grammar school *I acquired fair writing pretty soon, but I failed in the arithmetic, and made no progress in it* while later *being on some occasion made ashamed of my ignorance in figures, which I had twice failed in learning when at school, I took Cocker's book of arithmetic, and went through the whole by myself with great ease* [87]. He then used mathematics to alleviate boredom in the Pennsylvania assembly, recalling that he *was at length tired with sitting there to hear debates [...] which were so often unentertaining that I was induced to amuse myself with making magic squares or circles, or anything to avoid weariness* [87]. This is all that he says in his autobiography. Luckily, other people are more explicit, and Parton's *Life and Times of Benjamin Franklin* [222] explains that *The activity of Franklin's mind was shown in his trifling amusements. During the sessions of the Assembly, he had to endure many dull hours, perched in his seat as clerk, listening to debates in which he could take no part. His friend Logan showed him one day a French book of "Magical Squares," an idle game of the last century. Franklin, who had made these squares in his youth, now beguiled the tedium of the daily session*

S. Steinerberger, *The Unreasonable Elegance of Mathematics*,
Springer Undergraduate Mathematics Series, https://doi.org/10.1007/978-3-032-03815-9_15

by producing squares of extreme intricacy, surpassing all that had ever been done in that way.... [222]. Franklin constructed the following magical square.

200	217	232	249	8	25	40	57	72	89	104	121	136	153	168	185
58	39	26	7	250	231	218	199	186	167	154	135	122	103	90	71
198	219	230	251	6	27	38	59	70	91	102	123	134	155	166	187
60	37	28	5	252	229	220	197	188	165	156	133	124	101	92	69
201	216	233	248	9	24	41	56	73	88	105	120	137	152	169	184
55	42	23	10	247	234	215	202	183	170	151	138	119	106	87	74
203	214	235	246	11	22	43	54	75	86	107	118	139	150	171	182
53	44	21	12	245	236	213	204	181	172	149	140	117	108	85	76
205	212	237	244	13	20	45	52	77	84	109	116	141	148	173	180
51	46	19	14	243	238	211	206	179	174	147	142	115	110	83	78
207	210	239	242	15	18	47	50	79	82	111	114	143	146	175	178
49	48	17	16	241	240	209	208	177	176	145	144	113	112	81	80
196	221	228	253	4	29	36	61	68	93	100	125	132	157	164	189
62	35	30	3	254	227	222	195	190	163	158	131	126	99	94	67
194	223	226	255	2	31	34	63	66	95	98	127	130	159	162	191
64	33	32	1	256	225	224	193	192	161	160	129	128	97	96	65

Table 15.1: Benjamin Franklin's Magic Square of Magic Squares.

This is a magic square: the sum of the entries of every row and of every column is always the same number (2056). However, it's more magical than that: if we take any 4×4 sub-square, the sum of the entries is still 2056.

200	217	232	249
58	39	26	7
198	219	230	251
60	37	28	5

132	157	164	189
126	99	94	67
130	159	162	191
128	97	96	65

Table 15.2: The 4×4 sub-squares in the left top corner and bottom right corner: the sum of their entries, as for every other 4×4 sub-square, is also 2056.

Or in, the words of James Parton, *he made one of 16 cells in each row, which besides possessing the properties of the square given above (the amount, however added, being always 2056), had also this most remarkable peculiarity: a square hole being cut in a piece of paper of such a size as to take in and show through it just sixteen of the little squares, when laid on the greater square, the sum of the sixteen numbers, so appearing through the hole, wherever it was placed on the greater square, should likewise make 2056. This square was executed in a single evening. [...] Franklin himself jocularly said it was the 'most magically magical of any magic square ever made by any magician.'* [222]. For more details, see Pasles [224]. It is a curious coincidence that, when it comes to constructing 16×16 magic squares, Benjamin Franklin is in exquisite company: an entire section of Stifel's *Arithmetica Integra* is dedicated to constructing magic squares and, as he says, *In this matter, I will provide an example through which I will explain the entire concept. The example I will give results in each row summing to 2056.* However, a quick comparison shows that Franklin's magic square is a little bit more magical, Stifel's example does not have the property that each 4×4 subsquare sums to 2056.

Exemplum.

256	9	247	246	12	13	243	242	16	17	239	238	20	21	235	2
3	226	213	45	46	210	209	49	50	206	205	53	54	201	32	254
4	33	200	63	193	192	66	67	198	188	70	71	185	58	224	253
252	34	59	178	169	89	90	166	155	93	94	161	80	198	223	5
251	222	60	81	160	101	155	154	104	105	151	98	176	197	35	6
7	221	196	82	99	146	141	117	118	137	112	158	175	61	36	250
8	37	62	174	100	113	136	123	122	133	144	157	83	195	220	249
23	38	73	173	107	114	129	126	127	132	143	150	84	184	219	234
24	218	183	85	108	115	125	130	131	128	142	149	172	74	39	233
232	217	75	86	148	138	124	135	134	121	119	109	171	182	40	25
231	41	76	87	147	145	116	140	139	120	111	110	170	181	216	26
27	42	180	162	159	156	102	103	153	152	106	97	95	77	215	230
28	43	179	177	88	168	167	91	92	164	163	96	79	78	214	229
218	202	199	194	64	55	191	190	68	69	187	186	72	57	55	29
227	225	44	212	211	47	48	208	207	51	52	204	203	56	31	30
255	148	10	11	245	244	14	15	241	240	18	19	237	236	22	1

Fig. 15.1: Stifel's Magic Square [273].

Napoleon. Napoleon Bonaparte was known to be mathematically quite capable, and there is at least one result that bears his name. There is some debate whether it is really connected to Napoleon (see [118]), but it deserves to be mentioned.

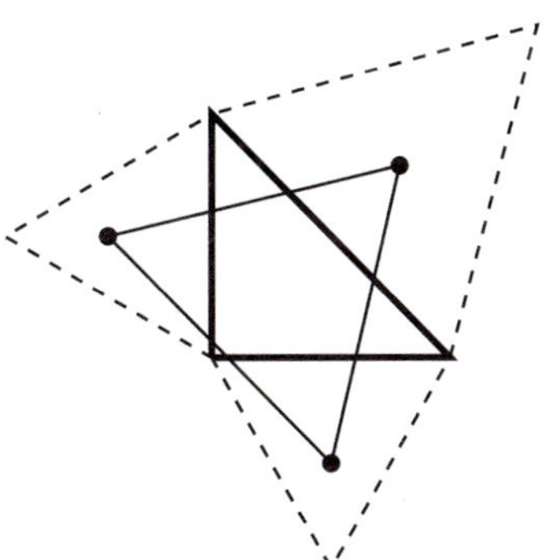

Fig. 15.2: Napoleon's Theorem: given any triangle (thick and solid), draw an equilateral triangle over each of the three sides (dashed). The triangle through the midpoints of the three equilateral triangles is also an equilateral triangle.

James Garfield. James Garfield (1831–1881) was the 20th President of the United States. In 1876, while being a Congressman for Ohio, he developed a *new* proof of the Pythagorean Theorem. The Pythagorean Theorem, surely one of the most famous results in all of mathematics has more than one proof: a book by Loomis [176] lists several hundred. Understandably, coming up with a new one is not so easy, but that is exactly what Garfield (possibly with some other members of Congress) did. His proof is listed as Proof No. 231 in [176].

PONS ASINORUM.

[In a personal interview with Gen. James A. Garfield, Member of
Congress from Ohio, we were shown the following demonstration
of the *pons asinorum*, which he had hit upon in some mathematical amusements and discussions with other M. C.'s. We do not
remember to have seen it before, and we think it something on
which the members of both houses can unite without distinction of
party.]

Fig. 15.3: The proof in the 1876 *New England Journal of Education* [104].

Thorold Gosset. John Herbert de Paz Thorold Gosset (1869–1962) was an English lawyer. As is described in Coxeter's *Regular Polytopes, He was called to the Bar in 1895, and took a law degree the following year. Then, having no clients, he amused himself by trying to find out what regular figures might exist in n dimensions. After rediscovering all of them, he proceeded to enumerate the 'semi-regular' figures. He recorded the results in the above-mentioned essay [...] That published statement remained unnoticed until after its results had been rediscovered by Elte and myself. As he was a modest man, Gosset let the subject drop, and pursued his career as a lawyer. He died in 1962 at the age of 93* [63]. Among all his discoveries is also a very regular structure in seven dimensions that is very difficult to visualize. However, one possible realization is shown in Fig. 15.4.

Fig. 15.4: The Gosset graph (visualization by Claudio Rocchini).

M. C. Escher. Maurits Cornelis Escher (1898–1972) was a Dutch graphic artist. He recalls that *I was an extremely poor pupil in arithmetic and algebra, and I still have great difficulty with the abstractions of figures and letters* [31]. This was confirmed by other people, his son recalling that *Father had difficulty comprehending that the working of his mind was akin to that of a mathematician* [64]. After finishing his studies at the Haarlem School for Architecture and Decorative Arts in 1922, he traveled to Italy and Spain. Visiting the *Alhambra*, one of the most famous examples of Islamic architecture in Spain, he encountered intricate tiling patterns that he studied and refined afterwards. This led to Escher acquiring a deep understanding of tiling structures. Even though he started to interact with mathematicians, he remained intellectually independent and discovered things mostly by himself. The mathematician H. M. S. Coxeter first encountered Escher's work at a 1954 conference. Escher asked Coxeter some mathematical questions to which he received short replies that he found difficult to comprehend. Escher's tenacity shines through:

> [Coxeter] encloses an example of using the values three and seven, of all things! However this odd seven is of no use to me at all; I long for two and four (or four and eight)... My great enthusiasm for this sort of picture and my tenacity in pursuing the study will perhaps lead to a satisfactory solution in the end. ... it seems to be very difficult for Coxeter to write intelligibly to a layman. Finally, no matter how difficult it is, I feel all the more satisfaction from solving a problem like this in my own bumbling fashion. (Escher [31])

For many more details and a much more extensive discussion, we refer to the article by Schattschneider [253] and the book by Coxeter, Emmer, Penrose and Teuber [64].

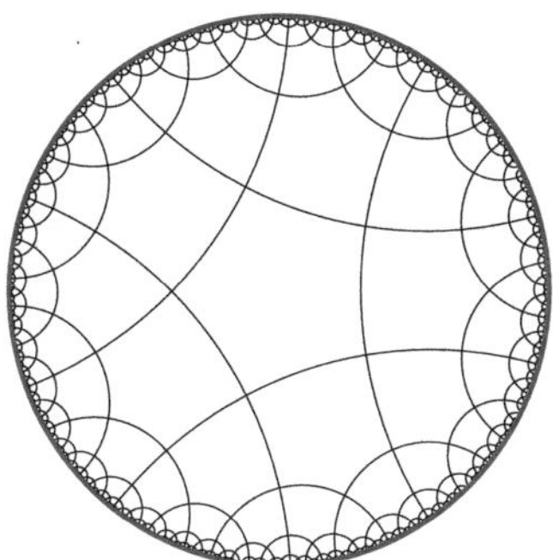

Fig. 15.5: The Poincaré Hyperbolic Disk can be found in some of Escher's work.

Edouard Zeckendorf. Edouard Zeckendorf (1901–1983) became a medical doctor in 1925, specializing in surgery and delivery. It is unclear when his interests turned towards mathematics: his first published result appeared in 1949, his last in 1978. The result that now carries his name ('Zeckendorf's Theorem') concerns the Fibonacci numbers $1, 2, 3, 5, 8, 13, 21, 34, 55, 89, \ldots$ that we have already seen before: each term is the sum of the preceding two terms.

Theorem (Zeckendorf [307], 1972). *Every positive integer can be represented uniquely as the sum of one or more distinct Fibonacci numbers in such a way that the sum does not include any two consecutive Fibonacci numbers.*

For example, picking the number 28, we see that we can write

$$28 = 21 + 5 + 2$$

and this happens to be the only way to write 28 as the sum of Fibonacci numbers in such a way that the involved numbers (here 2,5,21) are all Fibonacci numbers but also do not simultaneously show up next to each other in the Fibonacci sequence. Recalling Stigler's law that *No scientific discovery is named after its original discoverer* [274], it will not be a surprise that Zeckendorf's Theorem was first discovered by Lekkerkerker [170] almost 20 years earlier.

15.2 ...and colorful crackpots

Mathematics is fun, everyone can do it, many people enjoy it. Some end up getting a little bit carried away. Sometimes it happens that one thinks that one has discovered something new when this may not be the case. We all make mistakes, every mathematician has a dozen stories about having 'almost' solved a problem or rediscovering a known result. Two things separate the scientist from the crackpot.

1. The crackpot *knows* that their solution is correct: pointing out errors has no effect. The errors are always 'fixable', they are never crucial, the 'spirit' of the solution, we are told, is still correct.
2. The world *has* to know.

A hypothetical amateur scientist who has a disproof of the Pythagorean Theorem hidden away in a drawer would not qualify as a crackpot unless they are trying to (aggressively) tell people about it. Crackpots pose an interesting dilemma for the scientific community: many scientific breakthroughs seem too radical to be true, they seem too outlandish, they seem crazy. So if now someone claims to have some radical new mathematical theory, are they serious or are they crackpots? If we falsely assume that they are serious, we waste our time, potentially a lot of time! If we assume that they are crackpots when they are in fact revolutionary scientists, we risk going down in history as dullards who did not have the imagination to recognize a radical new advance when it is presented to them. This last scenario, experts not realizing revolutionary work when it is presented to them, happens fairly frequently. Suppose it is the year 1913 and you find yourself a Professor of Mathematics at Cambridge University and receive the following letter [26]

Dear Sir,
I beg to introduce myself to you as a clerk in the Accounts Department of the Port Trust Office at Madras on a salary of only £20 per annum. I am now about 23 years of age. I have

had no University education but I have undergone the ordinary school course [...] I have not trodden through the conventional regular course which is followed in a University course, but I am striking out a new path for myself [...] I have been developing this to a remarkable extent so much so that the local mathematicians are not able to understand me [...]

I remain, Dear Sir, Yours truly,
S. Ramanujan

Attached to the letter you find a list of very strange equations. *Some of the formulae were familiar and others seemed scarcely possible to believe. A few (concerning the distribution of primes) could be said to be definitely false* [124]. Do you take the letter seriously? Let us be honest, *I have no University education*, the notion of *striking out a new path* and doing this so remarkably that the *local mathematicians are not able to understand me*, do not inspire confidence – especially since some of the equations are quickly spotted to be wrong. Keep in mind that you probably receive several letters from over-enthusiastic amateurs a year, you have a lecture to prepare, and you are behind on other chores. The scenario is not at all fictional. The letter was sent to Prof. Henry Baker (who returned it without comment), Prof. Ernst Hobston (who returned it without comment), and Prof. Godfrey Hardy, who ended up taking it seriously. Hardy ended up bringing Ramanujan to Cambridge; the astonishing work done by Ramanujan (1887–1920) is nothing short of a miracle and continues to inspire to this day. Ramanujan's letters are not without humor

the sum of an infinite number of terms of the series:

$$1 + 2 + 3 + 4 + \cdots = -\frac{1}{12} \qquad \text{under my theory.}$$

If I tell you this you will at once point out to me the lunatic asylum as my goal. ([26])

Of course not everybody *striking out a new path* is a Ramanujan and this is exactly where the problem lies. The precise study of crackpottery may have begun with Fred Gruenberger and his article *A Measure for Crackpots: How does one distinguish between valid scientific work and counterfeit science?* [119, 120].

A MEASURE FOR CRACKPOTS

Fred J. Gruenberger*

The RAND Corporation, Santa Monica, California

Gruenberger starts with the observation that *For every article one sees in a technical journal or, for that matter, even in the public press, a decision has to be made: Is this worth reading or is it something that can safely be skipped?* He goes on to discuss the main problem

In a similar way it could be argued that a crackpot can be spotted by the (crackpot) way he tries to communicate. If he talks like a crackpot, fine; perhaps you can thereby damn him, but you take a real risk by doing so. Many weird ideas (weird, that is, after the acid test of time) have been advanced in the canonical form of true science. Yet there are many examples

> in history of people we now regard as outstanding scientists whose early writings look like
> those of a raving lunatic. (Gruenberger [120])

Gruenberger develops a quantitative criterion to test whether someone might be a crackpot. One beautiful aspect of the list is that it perfectly captures how a 'typical' crackpot behaves. The first few criteria are perhaps not surprising, the first being public verifiability. *The scientist says 'I did thus and so and observed its effect; you are free to repeat my steps.' The crackpot often says, 'This is revealed truth; sorry, but I and my followers are the only ones who can obtain these results'.* [120]. A very telling criterion is the tenth *This is a negative test. The true crackpot can frequently be spotted on this test alone. He proceeds with an argument like this: "They laughed at Fulton. He was right. They're laughing at me. Therefore, I must be an equal genius."* [120].

To make it concrete: the author has, over the years, received messages from people he never met who introduced themselves by saying that *as a genius mathematician who has been institutionalized 5 times, infinity weighs heavily on my soul.* One writer communicated that *As a long-term admirer of your work, I felt compelled to reach out to you after having come across a particular manuscript.* The 72 page pamphlet was conveniently attached, the email was sent to hundreds of people simultaneously. One email warned me *not to eat apples: their radiation is degrading people's mathematical abilities.* Some of these messages are not all that different from Ramanujan's clumsy first letter to Hardy. Which of these are worth one's time? To make matters more complicated, some of these messages were clearly written by people who are passionate about their work and filled with a strong desire to communicate it to the world. If I can spend five minutes of my time to brighten someone else's day, surely it is worth the time? On the other hand, these emails usually go to hundreds of people simultaneously, so why should I spend my time on them? Someone else will reply. The author recalls one incident outside the norm: a gigantic poster explaining a disproof of Einstein's Special Relativity Theory was delivered directly to the office, no letter attached. The attempted disproof was not easy to understand (there were no equations, only drawings). The author tried to find a motivated physics student to help but was unsuccessful.

15.3 Exercises

1. Try to prove Zeckendorf's theorem. Maybe you can even come up with a method that finds this unique representation?
2. Read Gruenberger's paper [120] and use it to compute the Crackpot Index for Losada's paper [179]. The Internet is filled with disproofs of Einstein's theory of Relativity (or anything else): compute their crackpot index.
3. To make matters more complicated, find a copy of Ramanujan's first letter and compute its crackpot index. Things are not always so simple.

Chapter 16
The House Always Wins

When from their game of dice men separate,
He, who hath lost, remains in sadness fix'd,
Revolving in his mind, what luckless throws
He cast (Dante, *Divine Comedy* [67])

16.1 The very early history: Gerolamo Cardano

The early history of probability theory is mysterious. While geometry, numbers, and shapes seem to show up as early as we have records, the same cannot be said of probability. The philosopher Ian Hacking points out that *Although dicing is one of the oldest of human pastimes, there is no known mathematics of randomness until the Renaissance. None of the explanations of this fact is compelling* [122]. One of the earliest works on the subject is by Gerolamo Cardano (1501–1576), who had *very* wide-ranging interests. Cardano wrote 200 books on all subjects including the books *On Water, On the Zeal of Socrates, Triceps* and a mysterious *Fourth Book on Hidden Knowledge.* He wrote a book *On Writing Books* and another one *On My Own Works.* He also composed four books *On the Urine*, his interest being deeply personal since, as he reports, *In 1536 I was overtaken with, it scarcely seems credible, an extraordinary discharge of urine; and although for nearly forty years I have been afflicted with this same trouble, giving from sixty to one hundred ounces in a single day, I live well* [45]. Cardano's interest in probability is also personal. In his autobiography, he recalls

> And since for many years – almost forty – I applied myself assiduously to gambling, it is not easy to say how greatly the status of my private affairs suffered, and with nothing to show for it. [...] Peradventure in no respect can I be deemed worthy of praise; for so surely as I was inordinately addicted to the chess-board and the dicing table, I know that I must rather be considered deserving of the severest censure. I gambled at both for many years, at chess more than forty years, at dice about twenty-five; and not only every year, but – I say it with shame – every day, and with the loss at once of thought, of substance, and of time. (Cardano, *The Book of My Life* [45])

One of Cardano's books carries the title *On Games of Chance.* Some have used this book to declare Cardano the founder of modern probability theory, others have used the very same book as a reason to argue that he missed the point; we examine both cases. Cardano starts his treatise by stating that gambling may be illegal *But in times of great anxiety and grief, it is considered to be not only allowable, but even beneficial. Also, it is permitted to men in prison, to those condemned to death, and to the sick, and therefore the law also permits it in times of grief. For certainly, if any occasion will justify it, none is so worthy of excuse as this one. In my own case,*

© The Author(s), under exclusive license to Springer Nature Switzerland AG 2025
S. Steinerberger, *The Unreasonable Elegance of Mathematics,*
Springer Undergraduate Mathematics Series, https://doi.org/10.1007/978-3-032-03815-9_16

when it seemed to me after a long illness that death was close at hand, I found no little solace in playing constantly at dice [46]. Then, in Chapter 11 (*On the Cast of Two Dice*), we start to find hints of real theory

> In the case of two dice, there are six throws with like faces, and fifteen combinations with unlike faces [...] As for the throws with unlike faces, they occur in pairs in the eighteen casts of equality, [...] But the throw (1,2) can turn up in two ways, so that for it there is equality in nine casts; and if it turns up more frequently or more rarely, that is a matter of luck. [46]

If we throw two dice, the result $\{1, 1\}$ requires both throws to be a 1, which happens with likelihood $1/36$. However, the result $\{1, 2\}$ can happen in two ways, by throwing a 1 and then a 2 or the other way around, and thus the likelihood is $2/36 = 1/18$. This looks like probability theory, however, it could also be argued that it is merely an exercise in enumeration. There are also hints of what we will later come to call 'expectation', the typical outcome. In assigning fair wagers, it is important, Cardano claims, to consider everything that can happen and how many possibilities there are for a favorable outcome.

> Other questions must be considered more subtly, since mathematicians also may be deceived, but in a different way. I have wished this matter not to lie hidden because many people, not understanding Aristotle, have been deceived, and with loss. So there is one general rule, namely, that we should consider the whole circuit, and the number of those casts which represents in how many ways the favorable result can occur, and compare that number to the remainder of the circuit, and according to that proportion should the mutual wagers be laid so that one may contend on equal terms. (Cardano, [46])

One of the reasons why people have argued *against* Cardano as the first person dealing with probability is Chapter 20 (*On Luck in Play*). Luck rules supreme, he writes, and promptly dismisses his own previous reasoning: *In these matters luck seems to play a very great role, so that some meet with unexpected success while others fail in what they might expect; so that the above reasoning about the mean does not apply.* While one may try to compute averages, *it is agreed by all that one man may be more fortunate than another, or even than himself at another time of life, not only in games but also in business, and with one man more than another and on one day more than another* and, in summary, *there is no rational knowledge of luck to be found in this, though it is necessarily luck.* Cardano even tries to outsmart luck by predicting the future but warns against it

> I have never seen an astrologer who was lucky at gambling, nor were those lucky who took their advice. The reason is as follows: although, as in all matters of chance, it happens that they occasionally make the right forecast, nevertheless, if they have guessed right (for when they do not guess right, they must lose) then immediately afterward they go very badly wrong, since they become venturesome and lose more from one mistake than they win from forecasting correctly four times in a row. For the path into error is always steeper, and the loss is greater than the gain. [46]

Chapter 20 continues, less philosophically, by detailing some of his own stories in which his true passion for gambling shines through. *Yet I have decided to submit to the judgment of my readers what happened to me in the year 1526 in the company of Thomas Lezius, the patrician of Venice, leaving it to each reader to form his own opinion. [...] We were playing a game (called Bassette) and I won all the money he had. [...] He promised to pay what he owed me within three days; but he did not come. Then he chanced to meet me and said that he would come to pay the money on Saturday (which was the day of the Nativity of the Virgin) and promised to take me to a beautiful prostitute. At that time I was just completing my twenty-fifth year, but I was impotent. Nevertheless, I accepted the condition...* – an interesting way for probability theory to begin! No discussion of Cardano would be complete if we did not mention that his father, Fazio Cardano (1444–1524), a professor at the University of Pavia and friend of Leonardo da Vinci, may have met extraterrestrials. We quote directly from Cardano's *De Subtilitate*, Chapter 19 (*'On Demons'*):

> But I append the most remarkable of all the tales, which I heard not just once, but a number of times from my father, Fazio Cardano, who used to claim that he had had a family demon for about thirty years. Finally, when I was looking into what he had written, I found what I had often heard set down in writing and for the record. *On the first day of August, 1491, when I had completed the religious rites at the twentieth hour of the day, seven men made their customary appearance, clad in silk clothing, with a robe like a Greek one, with purple boots (as it seemed), vests with shiny and reddish chest portions, so that they seemed to be from a chemise, of a shape narrower than the usual, and quite conspicuous. They were not all dressed like this, but two, who were accepted as the most important of them; there was another of them who was taller and fresh-faced, and two others were his followers; another who was rather pale and smaller of body; three others. So they were seven in all.* The written account left unstated whether there was anything on their heads or they were bare. Their age was around the fortieth year, but not less than the thirtieth. When asked who they were, they answered that they were, so to speak, people of the air, who were themselves born and died, but their life was much longer than ours, so that it might reach 300 years. On being questioned about the immortality of our soul, they stated that nothing survived of each person's own, they were themselves much closer to the gods than to the human race, but were still separated from them by an almost infinite gap. In comparison to us, that they are either happier or more wretched, as we are in comparison to the beasts. Nothing of the hidden things is concealed from them, for instance not books, nor money; and their lowest refuse is the geniuses of the most important men, just as the most worthless of people are the trainers of superior dogs or horses. They said that since they are themselves very slim, they can bring us nothing to our advantage or detriment, apart from spectres and terrors – and then knowledge. [...] My father asked why they did not reveal treasures to humanity, if they knew about them? They replied that for anyone to pass this on to humanity was prohibited by the private law, under extreme penalties. They stayed with him for more than three hours [...] (Cardano, *De Subtilitate*, [48])

16.2 Early History

The history of 'serious' probability theory starts suddenly and abruptly around 1660. In 1657, Christiaan Huygens invented both the pendulum clock and published *De Ratiociniis in Ludo Aleae* (On reasoning in random games) which starts

> Although in games depending entirely upon Fortune, the Success is always uncertain; yet it may be exactly determin'd at the same time, how much more likely one is to win than lose. (Huygens [143])

More gambling! Around the same time, in 1662, John Graunt published *Natural and Political Observations Made Upon the Bills of Mortality*. He used the *Bills of Mortality*, weekly published documents about the number of births, deaths and causes of deaths, to estimate the number of people living in London. Also in 1662, Antoine Arnauld and Pierre Nicole publish, possibly with the help of Blaise Pascal, *La logique, ou l'art de penser* (Logic, or the Art of Thinking), which contains, in Chapter 16 ('Of the Judgement which we should make touching future events'), paragraphs such as

> It is this which attracts so many to lotteries. Is it not a most advantageous thing, say they, to gain twenty thousand crowns for a single crown? Each believes that he is the happy one to whom the prize will fall; and no one reflects, that if there be, for example, twenty thousand crowns, it will be, perhaps, thirty thousand times more probable that each individual will lose them, than that he will gain them. (Logic, the Art of Thinking [11])

At the end of *La logique, ou l'art de penser*, on the very last page, our entire existence is turned into a gamble: since our afterlife will be infinite and our life is finite *This is why the smallest degree of facility for the attainment of salvation is of higher value than all the blessings of the world put together; and why the slightest peril of being lost is more serious than all temporal evils, considered simply as evils. This is enough to lead all reasonable persons to come to this conclusion, with which we will finish this Logic: that the greatest of all follies is to employ our time and our life in anything else but that which will enable us to acquire one which will never end, since all the blessings and evils of this life are nothing in comparison with those of another; and since the danger of falling into these evils, as well as the difficulty of acquiring these blessings, is very great* [11]. Leibniz also added to the mix: he considered the idea of assigning a value between 0 and 1 to statements with 0 meaning impossibility, 1 meaning certainty and everything else, well, being somewhere in the middle. More precisely, Leibniz *interpreted the four modalities by means of a probability calculus. We have to keep in mind that in 1669 he probably knew nothing of the works of Blaise Pascal or Pierre de Fermat on probability. Nevertheless, in the* Doctrina conditionum, *he develops the idea of representing the modalities necessarium by 1, impossibile by 0 and contingent/possibile by a fraction between 1 and 0* [10]. Somehow, suddenly around 1660, everything changes. The philosophically inclined reader will see two different sides of probability: one is about the behavior of dice as in Huygens or the outcome of a lottery, a probability that *is statistical, concerning itself with stochastic laws of chance processes* [122]. In contrast, the argument involving the afterlife is more a question of quantifying one's own personal *belief*. The same is true for Leibniz who was *dedicated to assessing reasonable degrees of belief in propositions quite devoid of statistical background* [122]. This fundamental duality is with us to this day. Consider the statement

> The likelihood of rain tomorrow is 70%.

A sentence like this is frequently found on TV, apps, the weather service, the newspaper and so on. What does it really mean? It is either going to rain or not going to rain, what are the 70%? In the dual sense of probability sketched above, two interpretations come to mind. The first would be that, when checking our records for days at a similar time of the year with similar weather conditions as we find today, in 7 out of 10 cases it rained the next day. The second case may perhaps be understood as a quantification of personal belief: I believe that the chance of rain tomorrow is 70%, I believe it is more likely than not, and if you were to offer me 1:1 odds, I would take that bet. This duality of interpretation is well-understood, and it is confusing that we use *probability* to refer to both of them. Carnap wanted to call them probability$_1$ and probability$_2$, Poisson and Cournot suggested to use *chance* (luck) and *probabilité* (probability) and Condorcet suggested *facilité* and *motif de croire*. Today, sometimes the terms *frequentist* and *Bayesian* are being used. We shall leave the philosophical aspects of probability aside and refer to Hacking [122] for more details about the adventurous story of early probability theory.

16.3 Random variables

Time to consider a random variable X. One may think of X as a die (not a threat: die is the singular of dice) or a coin or something else that gives a random value each time it is being evaluated. We restrict ourselves to random variables that only attain finitely many different real values such as dice or coins. Consider a simple thought experiment: someone offers us the following game.

> **Game.** You pay x. In exchange you get to throw a single die and get the amount indicated by the die. The die is fair and assumes each of the value $1, 2, 3, 4, 5, 6$ with likelihood $1/6$.

For which value of x would you pay the game? It is clear that if $x = 1$, it pays off to play (one cannot lose) and for $x = 6$ you would never play (one cannot win). A little more investigation naturally leads to the notion of the expected value. This notion is, in some sense, already contained in Huygens 1657 *De Ratiociniis in Ludo Aleae*, where he states, at the very beginning *As a Foundation to the following Proposition, I shall take Leave to lay down this Self-evident Truth: That any one Chance or Expectation to win any thing is worth just such a Sum, as wou'd procure in the same Chance and Expectation at a fair Lay. As for Example, if any one shou'd put 3 Shillings in one Hand, without telling me know which, and 7 in the other, and give me Choice of either of them; I say, it is the same thing as if he shou'd give me 5 Shillings; because with 5 Shillings I can, at a fair Lay, procure the same even Chance or Expectation to win 3 or 7 Shillings* [143]. Tracing down a couple more cases, we arrive at the following definition.

> **Definition.** Suppose a random variable X assumes the values $x_1, \ldots, x_n$ with likelihoods $p_1, \ldots, p_n$. Then the expected value of X, written $\mathbb{E}X$, is defined as
>
> $$\mathbb{E}X = p_1 x_1 + p_2 x_2 + \cdots + p_n x_n.$$

In the case of a die, where each of the numbers $1, 2, 3, 4, 5, 6$ is attained with equal likelihood $1/6$, we have

$$\mathbb{E}X = \frac{1}{6} \cdot 1 + \frac{1}{6} \cdot 2 + \cdots + \frac{1}{6} \cdot 6 = 3.5.$$

Since this is the typical value of a die, maybe we would be inclined to play the game when $x < 3.5$ and less inclined to play it when $x > 3.5$. This can be tested. Running a computer to play 1000 rounds of the game and reporting the average payoff per round, we see a behavior as shown in Fig. 16.1.

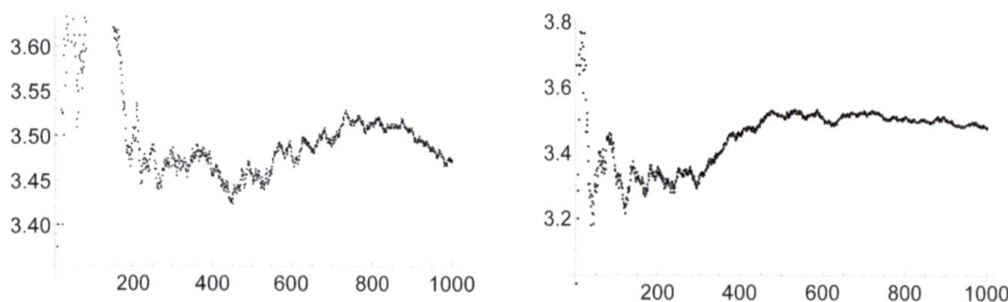

Fig. 16.1: Evolution of average outcomes when playing 1000 games.

Even though everything is random, a pattern yet emerges. In the first game, we are unexpectedly lucky in the first 200 steps of the game before then suddenly becoming very unlucky and trending towards the average value of 3.5. The second game shows a similar pattern, in both cases the long-time behavior is given by the expected value.

Fig. 16.2: Jakob Bernoulli's 1713 masterpiece: *The Art of Conjecturing* by Jacob Bernoulli, Professor in Basel and *very famous mathematician*.

This is one of the most central results in probability theory and is usually attributed to Jakob Bernoulli (1655–1705), who discusses it in his 1713 book *The Art of Conjecturing* (Ars Conjectandi) published posthumously by his nephew Nicolas. Bernoulli describes the main idea already 10 years earlier in a letter to Leibniz in which he wrote *Even though it is, through some form of natural instinct, obvious to even the dumbest among us that more and more observations make it less likely for us to miss our target, it is nonetheless not easy to accurately prove this* (J. Bernoulli to Leibniz [157]). It took Bernoulli a long time to make these ideas precise. We, like Bernoulli, will be imprecise at first, a more precise formulation is given later.

Theorem (Law of Large Numbers). *If X is a random variable and we evaluate it n times to get $X_1, \ldots, X_n$, then the average*

$$\frac{X_1 + X_2 + \cdots + X_n}{n}$$

is very likely to be close to the expected value $\mathbb{E}X$ (the larger n is, the more likely).

One may think of the expected value as the first fundamental piece of information when it comes to a random process. The Law of Large Numbers tells us what we can expect if we keep playing the random game forever. It is also an explanation of Fig. 16.1: as we keep throwing dice, the average outcome will, on average, get closer and closer to the expected value 3.5. Another beautiful example is given by slot machines in a casino: you pay a fixed amount of money (say 1 dollar) to play, the machine lights up and you get a random amount of money back: often nothing but sometimes a lot of money. The author will confess that when he first tried a slot machine (*Da Vinci Diamonds*), 1 dollar turned into a 27 dollar payout, which caused a lot of excitement. However, the Law of Large Numbers always wins and the author now avoids casinos. Casino slot machines come with a RTP ('Return to Player') number that corresponds to the expected value: the RTP describes the expected amount of dollars that are returned to the player when the player spends 100 dollars. The beloved *Da Vinci Diamonds* has an RTP of 94.94, which means that for every 100 dollars spent on this wonderful device, one loses, on average, 5.06 dollars. Various machines have an RTP as high as 98 or 99 and if you are forced to gamble in a casino, these are your friends – though if you are not forced to gamble, perhaps you should not; this is also reflected in the title of a classic textbook by Dubins and Savage, *How to Gamble If You Must: Inequalities for Stochastic Processes* [73].

16.4 Some psychological considerations

Does the expected value really tell the full story? Maybe not! One of the first people to realize this was the Swiss mathematician Daniel Bernoulli (1700–1782). The Bernoulli family has produced several great mathematicians, including Jakob Bernoulli (1654–1705), his brother Johann Bernoulli (1667–1748), their nephew Nicolas Bernoulli (1687–1759) and said Daniel, son of Johann (several others have been omitted for brevity). Daniel Bernoulli, in 1738, writes in his *Exposition of a New Theory on the Measurement of Risk* a paragraph that is shockingly modern and deserves to be read in full.

> To make this clear it is perhaps advisable to consider the following example: Somehow a very poor fellow obtains a lottery ticket that will yield with equal probability either nothing or twenty thousand ducats. Will this man evaluate his chance of winning at ten thousand ducats? Would he not be ill-advised to sell this lottery ticket for nine thousand ducats? To me it seems that the answer is in the negative. On the other hand I am inclined to believe that a rich man would be ill-advised to refuse to buy the lottery ticket for nine thousand ducats. If I am not wrong then it seems clear that all men cannot use the same rule to evaluate the gamble. The rule established in §1 must, therefore, be discarded. But anyone who considers

the problem with perspicacity and interest will ascertain that the concept of value which we have used in this rule may be defined in a way which renders the entire procedure universally acceptable without reservation. To do this the determination of the value of an item must not be based on its price, but rather on the utility it yields. The price of the item is dependent only on the thing itself and is equal for everyone; the utility, however, is dependent on the particular circumstances of the person making the estimate. Thus there is no doubt that a gain of one thousand ducats is more significant to a pauper than to a rich man though both gain the same amount. (Daniel Bernoulli, [27])

In modern language, we are dealing with a lottery ticket that is worth either 0 or 20,000 ducats, both equal likelihood. Its expected value is 10,000 ducats. Suppose I am now presented with two options: cash the lottery ticket and see whether I got lucky or sell the ticket for 9,000 ducats. What am I going to do? What *should* I do? Bernoulli argues, convincingly, that if I am very poor and if 9,000 ducats were to change my life drastically for the better, it might be a good idea to sell the ticket. Bernoulli also gives a convincing explanation as to why this might be the case.

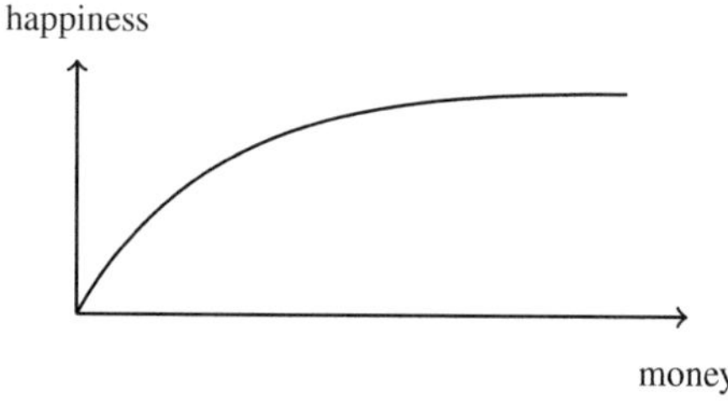

Fig. 16.3: Happiness may well be a nonlinear function of money.

People do not necessarily try to maximize the amount of ducats they have, they optimize for some sort of happiness. There are plenty of studies trying to connect happiness to income, and they paint a coherent picture that is sketched in Fig. 16.3. Having more money does indeed make you happier, however, having twice as much money does *not* make you twice as happy. What one observes is that when you are poor, every single ducat makes a difference; being in poverty is a challenging way to be alive, and escaping poverty has a significant impact on one's happiness. As one accumulates more wealth, each individual ducat makes less of a difference. Someone whose salary goes from 1000 ducats to 2000 ducats will experience significantly more happiness than someone whose salary goes from 10,000 ducats to 11,000 ducats. The technical term of this is *concavity of the utility function*. With this in mind, we can revisit Bernoulli's problem. Our choice is to get 0 or 20,000 ducats with likelihood 1/2 each or to get 9,000 ducats with certainty. If our happiness function had the values

$$h(0) = 0 \qquad h(9000) = 9 \qquad \text{and} \qquad h(20000) = 17$$

then our choice is between a guaranteed happiness of $h(9000) = 9$ and an expected happiness of

$$\frac{1}{2}h(0) + \frac{1}{2}h(20000) = \frac{17}{2} = 8.5 < 9$$

and it makes sense to sell the ticket.

16.5 The Allais Paradox

We have already been convinced, by Daniel Bernoulli, that humans do not necessarily maximize money. The French physicist and economist Maurice Allais (1911–2010) convincingly argued that we also do not maximize happiness (or utility) – this goes against a foundational assumption in modern economics and is a passionate matter. Allais writes that *It is particularly impressive to see that a mind as distinguished as Mr. Samuelson, after initially taking an attitude opposed to the central thesis of the Theory of Games, has finally accepted it two years ago and that he defends it today with great energy* [4]. (Conversely, Samuelson said of Allais *Had his earliest writings been in English, a generation of economic theory would have taken a different course.*) Allais, in the true spirit of a versatile scientist, also reported that pendulums may behave slightly differently during a solar eclipse; now known as the *Allais effect*, it may not exist. The Allais paradox, on the other hand, most definitely exists: you are given a choice between two options

$$\text{you get } 2400 \quad \text{or} \quad \begin{cases} 2500 & \text{with likelihood } 0.33 \\ 2400 & \text{with likelihood } 0.66 \\ 0 & \text{with likelihood } 0.01 \end{cases}$$

We see quickly that the second choice offers a slightly higher expected value since

$$2500 \cdot 0.33 + 2400 \cdot 0.66 + 0 \cdot 0.01 = 2409.$$

However, it also opens the door to receiving absolutely nothing; it is not tremendously likely (just 1%) but it can happen. Would you really be willing to risk the chance? The author would not, the author would go for the guaranteed 2400. The psychologists Kahneman and Tversky [147] report that in a study undertaken with 72 people, an overwhelming majority (82%) went with the guaranteed 2400. Let us now consider another problem, your two options are

$$\begin{cases} 2500 \text{ with likelihood } 0.33 \\ 0 \text{ with likelihood } 0.67 \end{cases} \quad \text{or} \quad \begin{cases} 2400 & \text{with likelihood } 0.34 \\ 0 & \text{with likelihood } 0.66. \end{cases}$$

These two options seem quite comparable. Their relative expected values are

$$2500 \cdot 0.33 = 825 > 816 = 2400 \cdot 0.34.$$

Most participants in the study (again around 82%) went again with the first option (and so would the author). If you find both these choices to be reasonable, you run into an interesting paradox. As we already established above, we are not trying to maximize for money, we are trying to maximize for happiness. I know that having no money brings me no happiness $h(0) = 0$. While I do not have a very precise understanding of what levels of happiness 2400 or 2500 ducats would get me, the fact that I prefer the safe 2400 option in the first game tells me that

$$h(2400) > 0.33 \cdot h(2500) + 0.66 \cdot h(2400).$$

Conversely, my choice in the second game tells me that

$$0.33 \cdot h(2500) > 0.34 \cdot h(2400) \quad \text{or, equivalently,} \quad h(2500) > \frac{0.34}{0.33} \cdot h(2400).$$

Plugging the second equation back into the first now gives us

$$h(2400) > 0.33 \cdot h(2500) + 0.66 \cdot h(2400)$$
$$> 0.33 \frac{0.34}{0.33} \cdot h(2400) + 0.66 \cdot h(2400) = h(2400).$$

However, $h(2400) > h(2400)$ is not true and something must go wrong: the way we choose between the two options in the game above is clearly not guided by maximizing money but, as this argument shows, is also not guided by maximizing happiness. The story is more complicated: the pain of losing 100 ducats exceeds the joy of winning 100 ducats. In the first scenario, we vividly feel the 1% chance of ending up with nothing and the regret we would experience. The Allais paradox has tremendous and highly nontrivial implications: one of these implications is in the medical sciences. Enter Kahneman and Tversky [283]: *The modern theory of decision making under risk emerged from a logical analysis of games of chance* and they argue that this is an incorrect way to describe actual human behavior: *deviations of actual behavior from the normative model are too widespread to be ignored, too systematic to be dismissed as random error, and too fundamental to be accommodated by relaxing the normative system.* They provide an illuminating example. Suppose you are facing lung cancer and have to make the difficult choice between surgery or a radiation treatment. The facts are as follows.

1. Surgery: Of 100 people having surgery, 90 live through the postoperative period, 68 are alive at the end of the first year, and 34 are alive at the end of five years.
2. Radiation: Of 100 people having radiation therapy, all live through the treatment, 77 are alive at the end of one year, and 22 are alive at the end of five years.

It is a difficult choice. In a survey of 247 people, 82% decided to go with the surgery. Suppose the choice is now presented differently.

1. Surgery: Of 100 people having surgery, 10 die during surgery or the post-operative period, 32 die by the end of the first year, and 66 die by the end of five years

2. Radiation: Of 100 people having radiation therapy, none die during treatment, 23 die by the end of one year, and 78 die by the end of five years.

It is still a difficult choice because *it is the exact same choice as above*. However, in a survey of 336 people exposed to this formulation, only 56% decided to go with the surgery. Again, loss aversion plays a significant role. *It is easier to forgo a discount than to accept a surcharge because the same price difference is valued as a gain in the former case and as a loss in the latter. Indeed, the credit card lobby is said to insist that any price difference between cash and card purchases should be labeled a cash discount rather than a credit surcharge* [283].

16.6 The Saint Petersburg Paradox

Even the idea of the expected value as 'the typical outcome' or the 'fair price of a game' is not completely harmless. This is illustrated by a specific paradox invented by Nicolas Bernoulli who described it in a letter to Pierre Remond de Montmort. The letter is remarkable, it is very modest while simultaneously containing one of the most influential paradoxes in Probability Theory: *I have been prevented for some time from conducting new research on the subject of probability, which is why I have nothing to communicate to you. However, in return for the problems you proposed to me, whose solutions I will examine when I have some leisure, I will now present a few others, that deserve your attention [...] Although most of these problems are not difficult, you will find something quite interesting in them* [203]. A particularly nice description of the Saint Petersburg paradox was given by Nicolas's cousin Daniel.

> My most honorable cousin the celebrated Nicolas Bernoulli [...] once submitted five problems to the highly distinguished mathematician Montmort. [...] The last of these problems runs as follows: Peter tosses a coin and continues to do so until it should land "heads" when it comes to the ground. He agrees to give Paul one ducat if he gets "heads" on the very first throw, two ducats if he gets it on the second, four if on the third, eight if on the fourth, and so on, so that with each additional throw the number of ducats he must pay is doubled. Suppose we seek to determine the value of Paul's expectation. (D. Bernoulli, [27])

Phrased differently, we toss a coin until it shows up as "heads" for the first time. If the first toss is already "heads", then Paul gets 1 ducat. If the first toss is "tails" and the second is "heads", Paul gets 2 ducats. If the first two tosses are "tails" and the third throw is "heads", Paul gets 4 ducats. If the first three tosses are "tails" and the fourth is "heads", Paul gets 8 ducats and so on. Rephrasing it in a slightly more algebraic fashion, we end up with the following

> **Game.** You pay $x\$$. In exchange you get a single coin. You now get to toss the coin until you see heads for the first time. If it takes n tosses of the coin to see heads, you get $2^n\$$. For what values of x do you play the game?

Table de M. Pascal pour les combinaisons.

1.1.1.1.1.1.1.1.1. 1. 1. 1. 1.   1
1.2.3.4.5.6.7.8. 9. 10. 11. 12. 13
1.3.6.10.15.21.28. 36. 45. 55. 66. 78
1.4.10.20.35.56. 84.120.165.220. 286
1.5.15.35.70.126.210.330.495. 715
1.6.21.56.126.252.462.792.1287

Fig. 16.4: Montmort's 1713 book [203] discussing the 'Table of Monsieur Pascal for the combinations', the Pascal triangle!

What happens is the following: there's a 50% chance of the coin immediately coming up heads. Then we get 1\$. There's a 25% chance of first getting tails and then heads, okay, then we get 2\$. There's a 1/8 chance of first getting two tails and then a head, in that case we get 4\$. The expected value, the fair price of the game, is

$$\frac{1}{2} \cdot 1 + \frac{1}{4} \cdot 2 + \frac{1}{8} \cdot 4 + \cdots = \frac{1}{2} + \frac{1}{2} + \frac{1}{2} + \cdots = \infty.$$

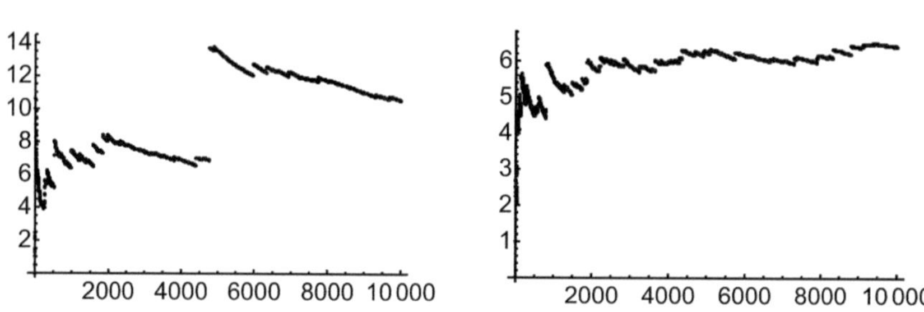

Fig. 16.5: The average outcome per round while playing the Saint Petersburg game for 10,000 rounds (left and right are two independent games).

Daniel Bernoulli continues saying *Although the standard calculation shows that the value of Paul's expectation is infinitely great, it has, he said, to be admitted that any fairly reasonable man would sell his chance, with great pleasure, for twenty ducats* [27]. The results of two simulations are shown in Fig. 16.5. In the first game our average winnings are initially close to 7 ducats until, around the 5000th game, we got unbelievably lucky and got 14 tails for a total win of $2^{15} = 32768$. The game is dominated by extreme events that occur very rarely (and none occur in the second game). In some sense, the expected value computation is correct: if I could play for a very, very long time, extremely rare events would eventually occur and I would ultimately be able to make money independently of how much it costs to play the game. However, the required timespan is astronomical and being able to play a very large number of times while having to pay a fee for each time actually implies that

I already need to have a lot of money to begin with. This is already observed by Bernoulli *this value will increase with the size of Paul's fortune and will never attain an infinite value unless Paul's wealth simultaneously becomes infinite. [...] From this we can easily see what a tremendous fortune a man must own for it to make sense for him to purchase Paul's opportunity for twenty* [27]. In summary, the secret behind the Saint Petersburg paradox is time: while the expected value still accurately measures the value of the game, the number of games is astronomical – and so is time and the required initial fortune.

16.7 Deviations from the expectation: variance

If we want to understand the behavior of a random variable, then the expected value is perhaps the most important piece of information. However, it by itself does not tell the full story. Consider game where we toss a coin and receive either 0 or 2 ducats (depending on the outcome of the coin toss) vs. simply receiving 1 ducat independently of what happens. In both cases, the expected value is 1, but these are very different scenarios. This was understood early on. The following paragraph in which Christiaan Huygens discusses life expectancy makes more sense if we recall that Nestor is an old hero in Homer's Iliad (*Nestor alone, improved by length of days* [140]) and that Methuselah is a biblical figure who lived for a *very* long time (*And all the days of Methuselah were nine hundred sixty and nine years* [28]).

> The quotient, 18 years and about 2.5 months, will be the value of the expectation of a newly conceived child. [...] Imagine that people were even feebler in their infancy than they are now, and that 90 in a 100 die before they are 6 but that those who exceed this age are veritable Nestors and Methusalehs, and they live on the average to be 152 years and 2 months. (Huygens, [122])

These numbers, 90% of people dying before they are age 6 with 10% living to be 152 years, are chosen so that the average again ends up being 18 years. We see three very different societies: (1) one where everyone dies at the age of 18, (2) the society around 1660, and (3) a society where most die very young and some live to be 152 years old. These three societies are very different, but their life expectancy is the same. The difference is clear: it is a question of how far away from the expected value $\mathbb{E}X$ we may expect the typical outcome to be. This leads to the notion of *variance*: if X is a random variable with expected value $\mathbb{E}X$, then the variance, denoted by $\mathbb{V}X$, is the typical value of the random variable $(X - \mathbb{E}X)^2$, formally

$$\mathbb{V}X = \mathbb{E}\left[(X - \mathbb{E}X)^2\right].$$

If a certain 'game' X always ends up paying us, without fail, one ducat, then $X = 1$ and $\mathbb{E}X = 1$ and $\mathbb{V}X = 0$. There is no variance, everything is deterministic. If, on the other hand, a random variable X assumes the values -1 and 1 with equal likelihood $1/2$, then $\mathbb{E}X = 0$ and

$$\mathbb{V}X = \mathbb{E}\left[(X - \mathbb{E}X)^2\right] = \mathbb{E}\left[X^2\right] = 1$$

since X^2 is always 1 independently of whether X is -1 or $+1$. As a last example, we compute the variance for a the throw of a single die X: the values $1, 2, 3, 4, 5, 6$ are all attained with equal likelihood $1/6$ and the expectation is 3.5, therefore

$$\begin{aligned}
\mathbb{V}X &= \mathbb{E}\left[(X - \mathbb{E}X)^2\right] \\
&= \frac{1}{6}(1 - 3.5)^2 + \frac{1}{6}(2 - 3.5)^2 + \frac{1}{6}(3 - 3.5)^2 \\
&+ \frac{1}{6}(4 - 3.5)^2 + \frac{1}{6}(5 - 3.5)^2 + \frac{1}{6}(6 - 3.5)^2 = \frac{35}{12} \sim 2.91\ldots
\end{aligned}$$

The history of variance is complicated. The special role played by a *sum of squares* shows up in the work of Carl Friedrich Gauss, particularly his *Method of Least Squares*. Gauss proposes to estimate quantities by ensuring that the sum of the squares of the error is minimized and writes in his *Theory of the Motion of the Heavenly Bodies moving about the sun in conic sections* that

> that will be the most probable system of values of the unknown quantities [...] in which the sum of the squares of the differences between the observed and computed values [...] is a minimum. This principle, which promises to be of most frequent use in all applications of the mathematics to natural philosophy, must, everywhere, be considered an axiom with the same propriety as the arithmetical mean of several observed values of the same quantity is adopted as the most probable value. (Gauss [106])

The method was first published by Legendre [166] in 1805. Gauss claimed to already have had the idea in 1795 which, naturally, caused a discussion over who actually discovered it. It is established that Gauss was able to use the method of least squares to spectacular effect in 1801. On the evening of Jan 1, 1801, the Italian astronomer Giuseppe Piazzi (1746–1826) observed what seemed to be a new star in the constellation Taurus. He checked again the next night and found that the relative position to a nearby star seemed to have changed. Stars do not change their relative position to each other, at least not within one night, and Piazzi was certain that he had made a mistake. However, measuring it again, he realized that it could not be a star since it was definitely moving relatively to the other stars. Piazzi's 'star' ended up moving closer to the sun, and became impossible to observe. Piazzi transmitted his existing observations to a number of other astronomers and proposed to name it Ceres Ferdinandea. The challenge was to predict the future path of Ceres based on what little information was available, and Gauss used the method of least squares to solve the problem. The astronomer Franz Xaver von Zach used Gauss' prediction to find Ceres on Dec 7, 1801, and was able to confirm its planetary nature on Dec 31, 1801. Another astronomer, Olbers, confirmed these results a day later [39, 86]. Ceres, moving with a diameter of nearly 1000km in the asteroid belt between the orbits of Mars and Jupiter, is now classified as a dwarf planet. It is difficult to imagine a more spectacular way to announce a new mathematical technique.

Even the planets confirm that the variance $\mathbb{V}X = \mathbb{E}\left[(X - \mathbb{E}X)^2\right]$ is a good idea. $\mathbb{V}X$ is the typical *squared* deviation from the expectation: this means that a random variable assuming values between 0 and 10 may well have a variance of 25 (which

Fig. 16.6: The only surviving portrait of Legendre is a *caricature* painted by Julien-Léopold Boilly. Make sure to sometimes take pictures of your loved ones!

happens, for example, if X assumes the values 0 and 10 with likelihood $1/2$). This may appear strange since 25 is much larger than the range of values assumed by the random variable. There is an easy fix: we can introduce the *standard deviation* by taking a square root, $\sigma = \sqrt{\mathbb{V}X}$. Why not simply work with the standard deviation to begin with? The reason is a remarkable fact by itself: if we have two random variables X, Y that are independent of each other (say, two independent coin tosses or one coin toss and one game of roulette), then the variance of their sum $X + Y$ is merely the sum of the variances: *this measure of uncertainty adds up.*

Theorem. *If X and Y are two random variables that are independent of each other,*

$$\mathbb{V}(X + Y) = \mathbb{V}(X) + \mathbb{V}(Y).$$

Proof. The proof is, mostly, by computation. We start by noting that

$$\mathbb{E}(X + Y) = (\mathbb{E}X) + (\mathbb{E}Y).$$

This is not too surprising: if I play roulette and simultaneously bet on horse races, the amount I can expect to win or lose is the sum of my expected returns. We write

$$\mathbb{V}(X + Y) = \mathbb{E}\left[((X + Y) - \mathbb{E}(X + Y))^2\right]$$
$$= \mathbb{E}\left[((X - \mathbb{E}(X)) + (Y - \mathbb{E}(Y))^2\right].$$

The binomial theorem $(a + b)^2 = a^2 + 2ab + b^2$ implies

$$((X - \mathbb{E}(X)) + (Y - \mathbb{E}(Y))^2 = (X - \mathbb{E}(X))^2$$
$$+ 2(X - \mathbb{E}(X))(Y - \mathbb{E}(Y)) + (Y - \mathbb{E}(Y))^2.$$

It remains to compute the expected value of this sum: as before, the expected value of a sum is the sum of the expected values, so we can perform these computations simultaneously. The expected value of the first expression is easy, by definition

$\mathbb{E}\left[(X - \mathbb{E}(X))^2\right] = \mathbb{V}(X)$ and, likewise, the third term turns into $\mathbb{V}(Y)$. It remains to deal with the mixed term. Here, we argue as follows: we think of $X - \mathbb{E}X$ as a game and of $Y - \mathbb{E}Y$ as a game. The typical payout of each game is 0 (because we subtract the fair price). Since the games are independent of each other, we expect

$$\mathbb{E}\left[(X - \mathbb{E}(X))(Y - \mathbb{E}(Y))\right] = 0$$

and this is indeed the case (see the exercises in 16.12). $\square$

This suggests thinking of a random variable X as an object that is, typically, of size $\mathbb{E}X$ and fluctuates at scale approximately $\sqrt{\mathbb{V}X}$. It will, typically, assume values that are roughly $\mathbb{E}X \pm \sqrt{\mathbb{V}X}$.

16.8 Markov, Chebychev and Random Sums

We will now introduce three of the main ideas in basic probability theory. They are so simple that we will be able to do this in four pages, and yet they are wonderful tools that tell us much about the behavior of a random variable X and how that is connected to the expected value $\mathbb{E}X$ and the variance $\mathbb{V}X$.

1. Markov. The first is Markov's inequality. It says that if X is a random variable that only assumes nonnegative values, meaning values ≥ 0, then it is unlikely to assume values that are much larger than $\mathbb{E}X$. Suppose, for example, there is a geography exam and the class *average* (on a scale from $0\% - 100\%$) is 12%. This does not tell us much about any individual student: it is possible that some students achieved a perfect score. However, it does imply that not all students can have scored higher than 80% (since the average would then be greater than 80%). Markov's inequality makes these notions precise.

Theorem (Markov's inequality). *If X is a random variable assuming only nonnegative values, $X \geq 0$, then, for all $t > 0$, the probability that $X \geq t$ satisfies*

$$\mathbb{P}\left(X \geq t\right) \leq \frac{\mathbb{E}X}{t}.$$

Proof. We assume the random variable X assumes the values $0 \leq x_1 \leq x_2 \leq \cdots \leq x_n$ with likelihoods $p_1, p_2, \ldots, p_n$. Note that $p_i \geq 0$ and $p_1 + \cdots + p_n = 1$. By definition of the expected value,

$$\mathbb{E}X = p_1 x_1 + p_2 x_2 + \cdots + p_n x_n.$$

If $t \leq x_1$, then X is always larger than t and $\mathbb{P}\left(X \geq t\right) = 1$. Likewise,

$$\frac{\mathbb{E}X}{t} = \frac{p_1 x_1 + p_2 x_2 + \cdots + p_n x_n}{t} \geq \frac{p_1 t + p_2 t + \cdots + p_n t}{t} = 1,$$

and the inequality is true. If $t > x_n$, then the inequality is surely true since $\mathbb{P}(X \geq t) = 0$. It remains to deal with the case where t is somewhere in between these values: we may assume that there exists $k \in \mathbb{N}$ such that

$$x_1 \leq x_2 \leq \cdots \leq x_k < t \leq x_{k+1} \leq \cdots \leq x_n.$$

In that case, we start with the definition of expectation, drop some terms (since the x_i are all positive) and then use the inequality $t \leq x_{k+1}$ to deduce

$$\begin{aligned}
\mathbb{E}X &= p_1 x_1 + p_2 x_2 + \cdots + p_n x_n \\
&\geq p_{k+1} x_{k+1} + p_{k+2} x_{k+2} + \cdots + p_n x_n \\
&\geq p_{k+1} t + p_{k+2} t + \ldots p_n t = t \cdot \mathbb{P}(X \geq t). \qquad \square
\end{aligned}$$

As above, let X denote the grade on the geography test of a randomly chosen student. Then, by setting $t = 100$, we deduce that the fraction of students that achieved a perfect score on an exam whose average score is 12% is bounded by

$$\mathbb{P}(X \geq 100) \leq \frac{\mathbb{E}X}{100} = \frac{12}{100} = 0.12$$

corresponding to at most 12% of the class. We see that this is in fact optimal: if 12% of the class write a perfect exam and the remainder of the class does not answer a single question correctly (resulting in 0% for this unlucky remainder), then the average score of the class is in fact 12%. We may also use the inequality to bound the number of students who got a score of at least 80% by setting $t = 80$. In that case, Markov's inequality tells us that in an exam with an average score of 12%, the fraction of students achieving a score of at least 80% cannot exceed

$$\mathbb{P}(X \geq 80) \leq \frac{\mathbb{E}X}{80} = \frac{12}{80} = 0.15$$

or 15% of the class. As will come as no surprise to the reader, the naming is not entirely accurate: Markov's inequality was already known to Chebychev (who was Markov's teacher) around the time of Markov's birth.

2. Chebychev. Just as Markov's inequality was already known to Chebychev, Chebychev's inequality was first formulated by the French mathematician Irénée-Jules Bienaymé (though Chebychev's formulation [54] was a little bit cleaner). Pafnuty Lvovich Chebyshev (1821–1894) is sometimes considered to be the founding father of Russian mathematics. His name is rather unique: the first name comes from Coptic ('Man of God') and his last name must hold the record in number of different ways it is spelled: Tchebichef, Tchebychev, Tchebycheff, Tschebyschev, Tschebyschef, Tschebyscheff, Cebycev, Cebysev, Chebysheff, Chebychov, Chebyshov and Chebychev have all been used at some point or another. A man of many names and many nice ideas, the 'Chebychev' inequality is one of them.

The Chebychev inequality is about the idea that a random variable X with small variance should usually assume values that are not too far from $\mathbb{E}X$.

CONSIDÉRATIONS

A l'appui de la découverte de Laplace sur la loi de probabilité
dans la méthode des moindres carrés;

Par M. BIENAYMÉ [*].

Fig. 16.7: (Monsieur) Bienaymé's 1853 paper *Considerations In Support of Laplace's Discovery on the Law of Probability in the Method of Least Squares* [29].

Theorem (Chebychev's Inequality). *The likelihood that a random variable X deviates from its expectation $\mathbb{E}X$ by more than $t > 0$ satisfies*

$$\mathbb{P}\left(|X - \mathbb{E}X| \geq t\right) \leq \frac{\mathbb{V}X}{t^2}.$$

Proof. The proof is short: think of $Y = (X - \mathbb{E}X)^2$ as a new random variable. It is non-negative, $Y \geq 0$, and $\mathbb{P}\left(|X - \mathbb{E}X| \geq t\right) = \mathbb{P}\left(Y \geq t^2\right)$. Markov's inequality implies $\mathbb{P}\left(Y \geq t^2\right) \leq (\mathbb{E}Y)/t^2$. However, $\mathbb{E}Y = \mathbb{E}(X - \mathbb{E}X)^2 = \mathbb{V}X$. $\qquad\square$

Random Sums. Suppose now that X is a random variable with expectation $\mathbb{E}X$ and variance $\mathbb{V}X$. We may think of this random variable as the random outcome of some game of chance. We now play the game n times and are presented with the random outcomes $X_1, \ldots, X_n$. A relevant quantity is the typical outcome, the average amount of money won

$$Y_n = \frac{X_1 + \cdots + X_n}{n}.$$

Bernoulli's law of large numbers suggests that this should be close to the expectation $\mathbb{E}X$. But how close is it? One answer is given by Chebychev's inequality, and this was already observed by Chebychev himself in 1867 when proving the (Bienaymé–)Chebychev inequality discussed in the previous section.

DES VALEURS MOYENNES;

Par M. P.–L. DE TCHÉBYCHEF.

Fig. 16.8: Chebychev's (here: Tchébychef's) 1867 paper [54] *On average values* about Chebychev's inequality and the law of large numbers.

Theorem (Law of Large Numbers). *Let X be a random variable with expectation $\mathbb{E}X$ and variance $\mathbb{V}X$. Then*

$$\mathbb{P}\left(\left|\frac{X_1 + \cdots + X_n}{n} - \mathbb{E}X\right| \geq t\right) \leq \frac{\mathbb{V}X}{n}\frac{1}{t^2}.$$

Even though this may look complicated, the proof is shockingly simple: apply Chebychev's inequality. Let us first start with a simple example: throwing dice. Then X assumes the values $1, 2, 3, 4, 5, 6$, the expectation is $\mathbb{E}X = 7/2$, and the variance is $\mathbb{V}X = 35/12$. Suppose we roll 1000 dice and compute the sum of these 1000 random numbers: we expect the result to be close to 3500: but how close? Using the inequality above, we get

$$\mathbb{P}\left(\left|\frac{X_1 + \cdots + X_{1000}}{1000} - 3.5\right| \geq t\right) \leq \frac{35}{12}\frac{1}{1000}\frac{1}{t^2} \sim \frac{0.003}{t^2}.$$

Plugging in $t = 0.1$, we see that the likelihood of the total sum being somewhere between 3400 and 3600 is at least 70%. Plugging in $t = 0.2$, we can compute that the likelihood of the sum being between 3300 and 3700 is at least 92.5%. This method really works, and it gives nice and explicit bounds. Moreover, if t is some fixed number and n becomes very, very large, then the probability of the average being outside of $[\mathbb{E}X - t, \mathbb{E}X + t]$ gets smaller and smaller and ultimately goes to 0. As a result, if we take the average of sufficiently many random numbers, the randomness cancels out, and we are going to end up very close to the expected value. This was already noted by Chebychev:

> When the number of trials becomes infinite, we obtain a probability, as close as we want to unity, that the difference between the [...] probabilities of this event [...] and the ratio of the number of repetitions of this event, to the total number of trials, is less than any given quantity. In the special case where the probability of the event remains the same during all trials, we have Bernoulli's theorem. (Chebychev [54])

We will now prove the Law of Large Numbers in this formulation: despite the importance (also philosophical importance) of the result, the proof is very short.

Proof of the Law of Large Numbers. We consider the new random variable $Y_n = (X_1 + \cdots + X_n)/n$. The proof can be summarized in one word: apply the Chebychev inequality to Y_n. We first observe that $\mathbb{E}Y_n = \mathbb{E}X$. The second observation concerns the variance $\mathbb{V}Y_n$: using that the variance of a sum of random variables is merely the sum of the variances, we arrive at

$$\begin{aligned}
\mathbb{V}Y_n &= \mathbb{E}\left(\frac{X_1 + \cdots + X_n}{n} - \mathbb{E}\frac{X_1 + \cdots + X_n}{n}\right)^2 \\
&= \frac{1}{n^2}\mathbb{E}\left((X_1 + \cdots + X_n) - \mathbb{E}(X_1 + \cdots + X_n)\right)^2 \\
&= \frac{1}{n^2}\mathbb{V}(X_1 + \cdots + X_n) = \frac{1}{n^2}(\mathbb{V}X + \mathbb{V}X + \ldots \mathbb{V}X) = \frac{\mathbb{V}X}{n}.
\end{aligned}$$

That's it: plugging this into Chebychev's inequality proves the result. $\qquad\square$

16.9 The Central Limit Theorem, Galton's Board, and Free Will

Consider a random variable X that is ± 1 with equal likelihood. This corresponds to a game in which I receive 1\$ if the coin comes up heads, and I have to pay 1\$ if it comes up tails. The expected value is $\mathbb{E}X = 0$. It is a 'fair' game, however, there are probably still some random fluctuations. Suppose now that we play n such games, then

$$\mathbb{V}(X_1 + \cdots + X_n) = \mathbb{V}X + \cdots + \mathbb{V}X = n.$$

This means $\mathbb{E}\left[(X_1 + \cdots + X_n)^2\right] = n$, and shows that we expect, typically, that $X_1 + X_2 + \cdots + X_n \sim \pm\sqrt{n}$. If we toss a coin 1000 times, we expect 500 heads and 500 tails, but it is not all that realistic that we get exactly 500 heads and 500 tails, we should really expect to see

$$500 \pm \sqrt{1000} \sim 500 \pm 31 \qquad \text{of each.}$$

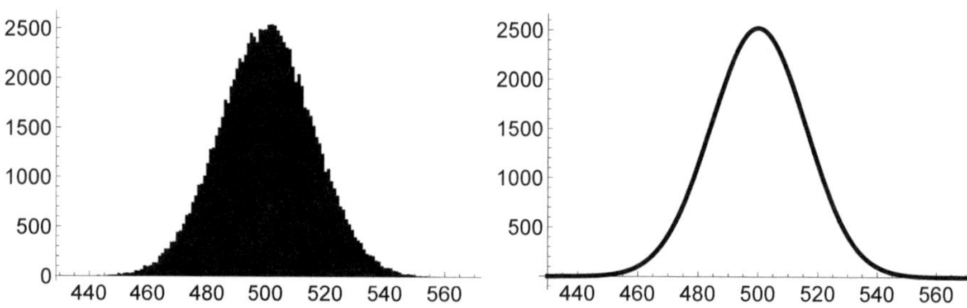

Fig. 16.9: Left: flipping 1000 coins and counting the number of heads, the outcomes of 100,000 such games sorted. Right: shape predicted by the Central Limit Theorem.

An experiment quickly confirms this: the results are shown in Fig. 16.9. Not only do we see that the prediction is accurate, we also see that there is also some type of 'curve' emerging. This curve represents what is arguably one of the most foundational, beautiful, surprising, and useful results in all of mathematics. The Central Limit Theorem makes the emergence of this curve precise. There are many versions of this result (see [85]), we give a simple informal description streamlined for clarity: we refer to the literature for the many different precise descriptions.

Theorem (Central Limit Theorem). *If $X = \pm 1$ with equal likelihood, then*

$$\frac{X_1 + X_2 + \cdots + X_n}{\sqrt{n}}$$

is a random variable. For n large, it behaves like a Gaussian random variable.

The standard Gaussian is a random variable that assumes continuous values with a likelihood given by the function $f(x) = \exp(-x^2/2)/\sqrt{2\pi}$. This formula is remarkable for a number of reasons: first, observe that it contains both π and Euler's number e. That, in itself, is fascinating. More important in practice is the shocking universality of this function: it shows up everywhere.

The Galton Board. Francis Galton (1822–1911) was an enthusiastic admirer of the Central Limit Theorem and came up with a particularly nice illustration: the Galton board. It is not well known by that name, but people tend to recognize it when they see it – all science museums sell one in their gift shop. Little balls enter at the top and then, when falling down, bounce randomly to the left or to the right. We may think of each random bounce as a ± 1 random variable (either left or right). The final position is going to be exactly $X_1 + X_2 + \cdots + X_n$ where n is the total number of bounces. The final resting position of the balls then mirrors the shape of a Gaussian random variable. So even though we do not know where any single individual ball is going to end up, if we observe *many* balls, then there is a strange sense of pre-determination.

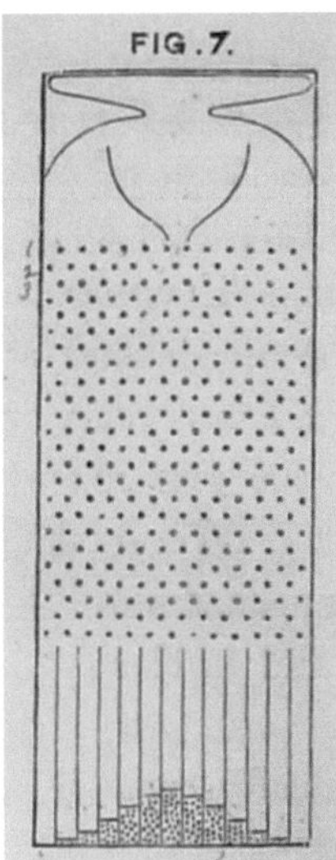

Order in Apparent Chaos. — I know of scarcely anything so apt to impress the imagination as the wonderful form of cosmic order expressed by the "Law of Frequency of Error." The law would have been personified by the Greeks and deified, if they had known of it. It reigns with serenity and in complete self-effacement amidst the wildest confusion. The huger the mob, and the greater the apparent anarchy, the more perfect is its sway. It is the supreme law of Unreason. Whenever a large sample of chaotic elements are taken in hand and marshalled in the order of their magnitude, an unsuspected and most beautiful form of regularity proves to have been latent all along. The tops of the marshalled row form a flowing curve of invariable proportions; and each element, as it is sorted into place, finds, as it were, a pre-ordained niche, accurately adapted to fit it. (Francis Galton, *Natural Inheritance* [101], 1894)

It is rare to hear of a mathematical theorem that *would have been personified by the Greeks and deified, if they had known of it*, but this is certainly a correct assessment of the strange beauty of the Central Limit Theorem. Francis Galton, a half-cousin of Charles Darwin, was an unusually versatile scientist. He invented the dog whistle, drew the first weather maps, worked on the structure of fingerprints, and coined the phrase 'nature versus nurture'. He also published a *Statistical Inquiries into the Efficacy of Prayer* [99] in which he investigates the question quite seriously: *The missionaries who are the most earnestly prayed for are usually those who sail on routes where there is little traffic, and therefore where there is more opportunity for the effects of secret providential overruling to display themselves than among those who sail in ordinary sea voyages.* [99]. It is hard to overestimate his originality.

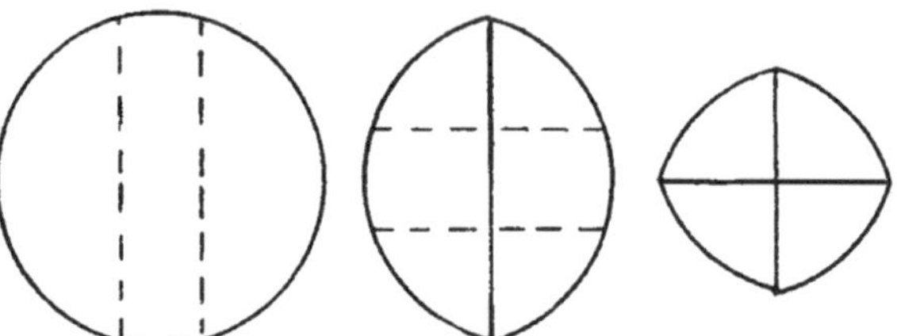

Broken straight lines show intended cuts. Ordinary straight lines show the cuts that have been made. The segments are kept in apposition by a common elastic band that encloses the whole. In the above figures about one-third of the area of the original disc is removed by each of the two successive operations.

Fig. 16.10: Galton's 1906 *Nature* cake article [102].

In 1906, he published the short article *Cutting a Round Cake on Scientific Principles* [102] that begins in a most enticing manner: *Christmas suggests cakes, and these the wish on my part to describe a method of cutting them that I have recently devised to my own amusement and satisfaction.* The main problem, he describes, is that the standard way of cutting cake leaves too much of the cake's surface out to dry, *the ordinary method of cutting out a wedge is very faulty in this respect* [102]. He proposes a radical new cake cutting technique (see Fig. 16.10) where middle segments are removed in such a way so as to leave the remainder symmetric. You eat the middle part, move the remaining parts together where *a [...] rubber band embraces the whole and keeps its segments together.* A proper British polymath, Galton also investigated

> THEORY OF TEA-MAKING. – I have made a number of experiments on the art of making good tea. [...] taking as my type of excellence, tea that was full bodied, full tasted, and in no way bitter or flat, I found that this was only produced when the water in the teapot had remained between 180° and 190° Fahr., and had stood eight minutes on the leaves. [...] There is no other mystery in the teapot. (Francis Galton [98], 1855)

Galton's 1855 *The art of travel, or, Shifts and contrivances available in wild countries* also investigates *Tea made in the kettle, Tea made in tin mugs, Tea made over night, Extract of Tea and Coffee*, and *Tea and Coffee, without hot water*.

Regression to the mean. Tea aside, one of the major discoveries of Galton was *Regression to the Mean*, which deserves its own quick comment. Galton mentioned in his 1886 *Regression towards mediocrity in hereditary structure* that *It is some years since I made an extensive series of experiments on the produce of seeds of different size but of the same species. They yielded results that seemed very noteworthy, and I used them as the basis of a lecture before the Royal Institution on February 9th, 1877. It appeared from these experiments that the offspring did not tend to resemble their parent seeds in size, but to be always more mediocre than they – to be smaller than the parents, if the parents were large; to be larger than the parents, if the parents were very small* (Galton [100]). This phenomenon is ubiquitous: random variables that take on extraordinarily large values are unlikely to take on as large values the next time around or, in the case of genetic factors determining height, are likely to only be partially transmitted to the next generation. The most fascinating example is perhaps the following: suppose you try to teach a difficult task to someone. They are practicing, and sometimes they do it better, sometimes they do it worse. When they do exceptionally well, they are unlikely to do better the next time around; conversely, when they do exceptionally badly, they tend to be better the next time.

> Danny [Kahneman] was then helping the Israeli Air Force to train fighter pilots. He'd noticed that the instructors believed that, in teaching men to fly jets, criticism was more powerful than praise. They'd explained to Danny that he only needed to see what happened after they praised a pilot for having performed especially well [...] The pilot who was praised always performed worse the next time out. (Michael Lewis, *The Undoing Project* [172])

Seems reasonable enough? *Danny watched for a bit and then explained to them what was actually going on: The Pilot who was praised because he had flown exceptionally well [...] simply [was] regressing to the mean. They'd have tended to perform [...] worse even if the teacher had said nothing at all* [172].

16.10 Free Will?

The Central Limit Theorem is a universal law arising out of chaos. If the individual balls in the Galton board all are forced to arrange along this Gaussian curve, where is the *free will* of each individual ball? Adolphe Quetelet (1796–1874) asked a similar question. He notes in his 1835 *On Man and the Development of his Faculties, or Essay on Social Physics* that

> It would appear, then, that moral phenomena, when observed on a great scale, are found to resemble physical phenomena; and we thus arrive, in inquiries of this kind, at the fundamental principle, that the greater the number of individuals observed, the more do individual peculiarities, whether physical or moral, become effaced, and leave in a prominent point of view the general facts, by virtue of which society exists and is preserved. (Quetelet [238])

This, however, causes some serious problems: if there are statistical regularities that govern our collective behavior, how can we truly be free? Quetelet continues and says *the effect of free will declines and disappears when observations involve a large number of people. All individual actions cancel each other out and revert to the class of effects produced by purely random causes* [260]. Quetelet continued and looked at marriages that involve a groom below the age of 30 and analyzed their number by the age of the bride. The result is shown in Table 16.1.

Ask yourself: if each year somewhere between 5 and 7 Belgian men below the age of 30 decide to marry a woman above the age of 60, do any of us really control our actions? It feels unlikely, too choreographed, something else must be pulling the strings behind the scene. Quetelet is deeply worried by Table 16.1: *We do not know, certainly, any statistical document more curious nor more instructive at the same time. To see almost identical numbers occurring from year to year one can never believe that chance is responsible for parallel arrangements; something mysterious is

Age of Bride	1841	1842	1843	1844	1845
≤ 30	12,788	12,422	12,368	13,024	13,157
(30, 45]	2,630	2,626	2,406	2,375	2,438
(45, 60]	93	121	125	129	102
> 60	7	6	8	5	5
Total Marriages	29,876	29,023	28,220	29,326	29,210

Table 16.1: Numbers of Belgian Marriages Classified by Age of Bride and Year of Marriage, for Bridegrooms of Age 30 and Under (Table taken from Seneta [260]).

happening beyond our understanding [260]. One possible explanation, he continues, is that we may have free will but only to a certain extent, *The Supreme Being has not accorded a power to man [free will] which tended to upset the laws governing all parts of creation; he has imposed limits on it just as he has to the ocean* [260]. Of course, there are many other interpretations along the lines of determinism through socialization. We are all social creatures; we have no choice over the first language we acquire, we have no choice regarding the culture or the circumstances we are born into. We usually do not choose the first schools we go to, and by the time we are able to make our own choices (are we ever?) we have experienced a lifetime of impressions not of our choosing.

> The empirical notion of freedom is: 'I am free if I can do what I want.' and through this 'what I want' the notion of freedom is determined. However, now that we are asking about the freedom of the desire, the question becomes [...] 'Can you also want what you want?' (Schopenhauer [256])

16.11 Bachelier and stock prices

The Gaussian distribution is absolutely wonderful: universal, easy to write down, easy to work with, and theoretically guaranteed by the Central Limit Theorem. A true marvel of science! However, there are limitations! This short story begins in

> En résumé, la considération des cours vrais permet d'énoncer ce principe fondamental :
>
> *L'espérance mathématique du spéculateur est nulle.*

Fig. 16.11: Bachelier's 1900 dissertation [15]: *In summary, the consideration of true prices allows us to state the following fundamental principle: The mathematical expectation of the speculator is 0.*

1900 when Louis Bachelier (1870–1946) published his dissertation *Théory de la spéculation* (The Theory of Speculation) in which he studied stock prices. Suppose the value of one stock of the *Bachelier Corporation* is worth 10$ today. It will have a different price tomorrow. If we knew already today that the stock price is going to be higher tomorrow, then I and many other people would have bought lots of stock and the price would already be higher today (the exception being insider trading). Likewise, if it is well-known that the stock price is lower tomorrow, we would have sold already. This means that the stock price will either go up or it will go down, but, in expectation, it will be 0. It is a natural assumption that the price of stock can be considered the sum of many small independent entities, and these would then sum up, according to the Central Limit Theorem, to a Gaussian.

Bachelier's Thesis. The fluctuations of stock market prices follows a Gaussian law.

Or, in the words of Bachelier, *One sees that the probability is governed by the Law of Gauss – already celebrated in the Theory of Probability* [15]. Bachelier's thesis introduces what we now call Brownian Motion (five years before Albert Einstein but 20 years after Thorvald Thiele's 1880 paper [278]). Brownian Motion is a process of random movement of particles with the property that the difference in position between any two points in time follows a Gaussian distribution. The process is named after the biologist Robert Brown (1773–1858), who observed pollen of a plant suspended in water under a microscope and noticed that they seemed to move in a random manner. The earliest description of such random movement is arguably due to the Roman philosopher Lucretius (99BC–55BC) in his masterpiece *De rerum natura* (On the Nature of Things). Lucretius' work is very modern: while not denying the existence of Gods, he argues that they do not interfere (*For all the gods must of themselves enjoy Immortal aeons and supreme repose, Withdrawn from our affairs, detached, afar*). He even argues that we should not be too concerned about what happens after death

> Look back:
> Nothing to us was all fore-passed eld
> Of time the eternal, ere we had a birth.
> And Nature holds this like a mirror up
> Of time-to-be when we are dead and gone.
> And what is there so horrible appears?
> Now what is there so sad about it all?
> Is't not serener far than any sleep?
> (Lucretius, *On the Nature of Things*, Book 3, [181])

Lucretius also observed a phenomenon that many of us are familiar with: if we find ourselves in a dark room with the curtain improperly closed, and if the sun is already high in the sky, there is often a very concentrated sunbeam entering the otherwise dark room. It is then easy to tell whether it is time to clean: dust particles are flying in the air. Lucretius has a more poetic description

> for behold whenever
> The sun's light and the rays, let in, pour down
> Across dark halls of houses: thou wilt see
> The many mites in many a manner mixed
> Amid a void in the very light of the rays,
> And battling on, as in eternal strife,
> (Lucretius, *On the Nature of Things*, Book 2, [181])

This movement of dust, observed by Lucretius, is the same that was observed by Robert Brown under the microscope and later suggested by Bachelier to describe the movement stock prices. A universal phenomenon?

Testing Bachelier's thesis. Bachelier makes a very concrete prediction: the difference between the price of a stock tomorrow and the price today is a random variable that follows a *Gaussian law*. This is a hypothesis that can be tested. Already in 1915, the American economist Wesley Mitchell notes

> While the actual and the "normal" distributions look much alike, they are not, strictly speaking, of the same type. The actual distribution is much more pointed than the other, and has a much higher "mode", or point of greatest density. (Mitchell [199])

A similar conclusion was reached by Mills [198] and Olivier [216]. In 1961, Alexander writes unambiguously that *A rigorous test [...] would lead us to strongly dismiss the hypothesis of normality* [3]. Mandelbrot [187] is unambiguous

> Despite the fundamental importance of Bachelier's process [...] it is now obvious that it does not account for the abundant data accumulated since 1900 by empirical economists, simply because *the empirical distributions of price changes are usually too "peaked" to be relative to samples from Gaussian population.* (Mandelbrot [187])

In 1973, Black and Scholes derive what is now known as the Black–Scholes formula: they are very precise in their assumptions and state that *In deriving our formula for the value of an option in terms of the price of the stock, we will assume "ideal conditions" [...] the distribution of possible stock prices at the end of any finite interval is log-normal* [30]. Avoiding a lengthy discussion about the difference between normal and log-normal, we note that Black and Scholes use an assumption that violates empirical evidence. Scholes was awarded the 1997 Nobel Prize in Economics. The Black–Scholes equation, or blind faith in that equation, is sometimes cited as a contributing element to the 2007–2008 financial crash. Mandelbrot, in his Autobiography [188], writes

> Thanks to the computer, I was able to note the flaws in Bachelier's model in a rough report I wrote in 1962, [...] The economics profession decided that my work was too complicated and too unfamiliar. The departure it represented and further threatened was hard to develop and sell. It seemed far easier to continue with an endless stream of "fixes." What was I to do? [...] And then, perhaps a bit later than I expected – in 2008 – the market did what it was bound to do: it crashed. (Mandelbrot [188])

16.12 Exercises

1. Ask people what they think *The likelihood of rain tomorrow is 70% really* means. See whether your friends are Frequentists (*probabilité*), Bayesians (*motif de croire*), or if they prefer not to speculate about the meaning of probability.
2. Roulette: you win 2\$ with likelihood 18/37, and you get 0\$ with likelihood 19/37. Compute the 'fair' price of the game, the expected value. How much money can the casino expect to make if you play 1000 thousand times (at cost 1\$)?

3. X and Y are independent random variables if the likelihood of X attaining a value a and, simultaneously, Y attaining the value b is merely the product of the individual likelihoods. For example, the likelihood of it raining tomorrow and my coin toss ending up "tails" is exactly half the likelihood of it raining tomorrow. Prove: if X and Y are independent, then $\mathbb{E}(X \cdot Y) = \mathbb{E}(X) \cdot \mathbb{E}(Y)$.

4. Use the Chebychev version of the Law of Large Numbers to bound the likelihood of 1000 rolls of a 20-sided die (assuming values $1, 2, \ldots, 20$ with likelihood 5% each) summing up to more than 13000.

5. Think about how much you would be willing to pay to play the Saint Petersburg Paradox game right now. I am assuming that you have a finite amount of money in your account and that you are not immortal. Explain your reasoning.

6. How does Quetelet's Marriage Table fit into your notion of free will?

7. Louis Bachelier claims that the difference between the stock price tomorrow and the stock price today is basically a random variable that behaves like a Gaussian. How would you go about disproving this? Get actual stock market data, and see what your methodology shows. Create an 'artifical' stock market price by adding random (gaussian) numbers (thus being as Bachelier predicted), and see whether you can detect a difference.

Chapter 17
Matching Things

17.1 Monge

We will now undertake a little excursion into the mathematical theory of 'how to move things around'. The theory of 'Optimal Transport' dates back to Gaspard Monge (1746–1818), who knew some interesting people.

> When on board the 'Orient' he [Napoleon] took pleasure in conversing frequently with Monge [...] Monge, endowed with an ardent imagination, without exactly possessing religious principles, had a kind of predisposition for religious ideas which harmonised with the notions of Bonaparte. (Bourrienne [34])

Monge is shown to be a widely interested traveler, *As there was no wood to be got, we collected a quantity of these bones for fuel. Monge himself was induced to sacrifice some of the curious skulls of animals which he had picked up on the way* [34]. He

Fig. 17.1: Gaspard Monge's 1781 *Mémoire sur la théorie des déblais et des remblais* in the Proceedings of the Royal Academy. Coincidentally, the article just before this one, from pages 657–665, is Borda's *Memoir sur les élections au scrutin* concerning election systems and introducing what we now know as the *Borda count*.

is even credited with explaining optical phenomena in the desert *On the 7th of July General Bonaparte left Alexandria for Damanhour. In the vast plains of Bohahire'h the mirage every moment presented to the eye wide sheets of water, while, as we advanced, we found nothing but barren ground full of deep cracks. [...] The cause of this singular illusion is now fully explained; and, from the observations of the learned Monge* [34] (see also [305]). Monge is buried in the Panthéon (possibly the most illustrious burial site in all of France, next to Voltaire, Rousseau, Lagrange, Victor Hugo, and Marie Curie). In 1781, Monge wrote his *Mémoire sur la théorie des déblais et des remblais* (Memoir about the Theory of Cut and Fill). This scientific paper is exceptionally well written. Loosely translated, it starts

S. Steinerberger, *The Unreasonable Elegance of Mathematics*,
Springer Undergraduate Mathematics Series, https://doi.org/10.1007/978-3-032-03815-9_17

When one has to transport soil from one place to another, it is customary to call the volume of soil that needs to be moved *déblai*, and the space it will occupy after transportation is called *remblai*. The cost of transporting a molecule, all else being equal, is proportional to its weight and the distance it must travel. Consequently, the total transportation cost must be proportional to the sum of the products of the molecules, each multiplied by the distance traveled. It follows that, given the shape and position of both the déblai and the remblai, it is not indifferent which molecule from the déblai is transported to which location in the remblai. There is a certain distribution of molecules between the two that minimizes the sum of these products, ensuring that the total transport cost is at a minimum. (Monge [201])

Let us make the problem even more concrete: suppose we live in a long narrow valley so that a position can be described with a single number $0 \leq x \leq 1$. Suppose there are n pizzerias, located in $x_1, \ldots, x_n$, these are n numbers in $[0, 1]$, and there are n houses, located in $y_1, \ldots, y_n$, also n numbers in $[0, 1]$. Each pizzeria has made a pizza, and each house wants exactly one pizza: sending a pizza from pizzeria x_i to house y_j comes with a certain delivery cost and this delivery cost is proportional to the distance $|x_i - y_j|$ between the pizzeria and the house; as said by Monge, *the cost of transporting a molecule* (or pizza) *is proportional to [...] the distance it must travel*. A solution of the problem is a function.

$$\sigma : \{1, 2, \ldots, n\} \to \{1, 2, \ldots, n\}.$$

Pizzeria 1 sends its pizza to the house with the number $\sigma(1)$, pizzeria 2 sends its pizza to house $\sigma(2)$ and so on. As part of the game, two different pizzerias $i \neq j$ have to send their pizza to two different houses $\sigma(i) \neq \sigma(j)$. This ensures that every house gets exactly one pizza. The problem is now to find a function σ so that

$$|x_1 - y_{\sigma(1)}| + |x_2 - y_{\sigma(2)}| + \cdots + |x_n - y_{\sigma(n)}| \qquad \text{is as small as possible.}$$

How do we find the optimal σ? We could try them all. This works well when n is small but, in general, requires us to check $n!$ (factorials!) different cases. If $n = 74$, and if we can convince each electron in the universe to compute one possible case for us each second, it will take the entire universe a year to check all the cases. Luckily for us, there exists a wonderful result that takes care of the problem: if we order both the pizzerias and the houses in increasing size, the optimal solution is easy.

Theorem. *Suppose the pizzerias and houses are ordered in increasing order*

$$x_1 \leq x_2 \leq \cdots \leq x_n \quad and \quad y_1 \leq x_2 \leq \cdots \leq y_n.$$

Then, for all one-to-one mappings $\sigma : \{1, 2, \ldots, n\} \to \{1, 2, \ldots, n\}$

$$|x_1 - y_{\sigma(1)}| + \cdots + |x_n - y_{\sigma(n)}| \geq |x_1 - y_1| + \cdots + |x_n - y_n|.$$

Put into words, the first pizzeria sends its pizza to the first house. The second pizzeria sends its pizza to the second house and so on. This seems pretty easy, so why is Monge's name on the Eiffel Tower? Suppose the pizzerias are located at $\{0.26, 0.38, 0.46, 0.64, 0.89\}$ and the houses at $\{0, 0.12, 0.2, 0.32, 0.87\}$. Applying the Theorem, we see that the pizzeria's should send their pizzas to the houses in the same increasing order

$$0.26 \to 0, 0.38 \to 0.12, 0.46 \to 0.2, 0.64 \to 0.32 \quad \text{and} \quad 0.89 \to 0.87.$$

The total distance traveled by all the pizzas is $|0.26 - 0| + |0.38 - 0.12| + \cdots + |0.89 - 0.87| = 1.12$ and this is an optimal solution. The first sign that this is nontrivial is that the solution is not unique. In the example above, $\sigma(i) = i$ for all $1 \le i \le 5$ is a solution. However, there exists a second solution that does just as well and is given by $\sigma(1) = 1, \sigma(2) = 3, \sigma(3) = 4, \sigma(4) = 2$ and $\sigma(5) = 5$.

Fig. 17.2: Delivering pizza! Left: the ordered solution that is guaranteed to be optimal. Right: another optimal solution.

One way to immediately end up in the middle of a big mystery is to critically reexamine the idea that transport cost is proportional to distance. Clearly, if we drive twice the distance, then this requires twice as much fuel. It takes twice as long and we have to pay the driver twice as much. However, there are some aspects where we are going to save some money: the shop that maintains our delivery trucks is going to cut us a better deal because we are bringing in twice the business. So maybe driving twice as far is not twice as expensive, maybe it is only 1.99 as expensive. Let us consider the possibility that, for some $0 < \alpha < 1$,

$$\text{the cost of transporting from } x_i \text{ to } y_j = |x_i - y_j|^\alpha.$$

As already observed by Wilfrid Gangbo and Robert McCann,

> the picture which emerges is rather different. Here the optimal maps will not be smooth, but display an intricate structure which – for us – was unexpected; it seems equally fascinating from the mathematical and the economic point of view. [...] To describe one effect in economic terms: the concavity of the cost function favors a long trip and a short trip over two trips of average length [...] it can be efficient for two trucks carrying the same commodity to pass each other traveling opposite directions on the highway: one truck must be a local supplier, the other on a longer haul. (Gangbo and McCann, 1996, [103])

Let us try to give a completely concrete example when the cost function is given by $c(x, y) = \log |x - y|$. This is similar to the example above when α is very close to 0. We pick 50 random points in $[0, 1]$ as the location of the pizzerias and another 50 random points in $[0, 1]$ as the location of the houses. One sees (see Fig. 17.3, left) a complicated structure emerging: most pizzerias transport to nearby houses, but some transport to houses that are *very* far away.

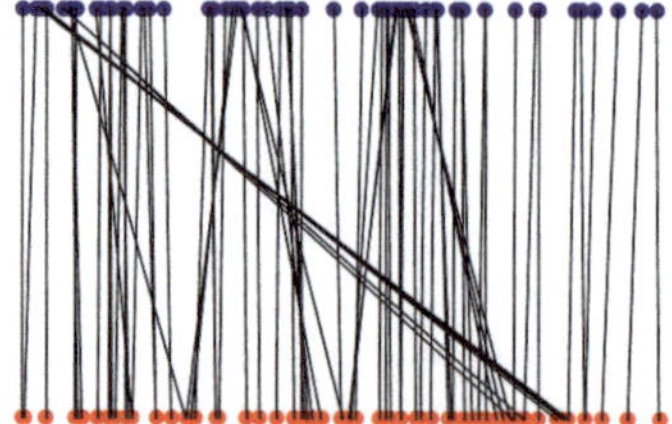 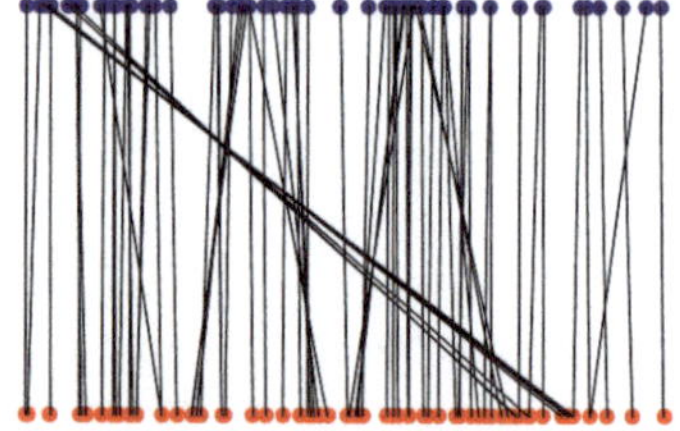

Fig. 17.3: Left: 50 red points on $\mathbb{R}$ being matched to 50 blue points on $\mathbb{R}$ when $c(x,y) = \log|x-y|$. Right: same points and same cost function matched with the greedy algorithm. The two matchings are very, very similar: why?

The story gets even weirder: there is a pretty simple way to assign pizzas to houses. Among all the pizzerias x_i and all the houses y_j, we could first try to find the pizzeria and the house that are the closest to each other among all pizzerias and all houses, the pair that minimizes $|x_i - y_j|$. Maybe that is the apartment just above the pizzeria; in any case, it is very close, and maybe the i-th pizzeria should deliver to the j-th house, we set $\sigma(i) = j$. Now we can remove the i-th pizzeria and the j-th house from consideration. We repeat the procedure and again look, among the remaining pizzerias and houses, for the ones with the smallest distance. Ultimately, we get an assignment σ of how to deliver pizzas. This procedure is commonly referred to as *greedy method*. It tries to do the best it can at each step with no consideration for the big picture, no long-term planning. Greedy methods are usually far from optimal. Here, however, we get results that seem to be close to optimal (see Fig. 17.3). In our setting, the greedy method has another fascinating property (see Fig. 17.4).

Theorem. *Suppose* $\{x_1, \ldots, x_n\} \subset [0,1]$ *are pizzerias and* $\{y_1, \ldots, y_n\} \subset [0,1]$ *are houses. For all* $1 \leq i \leq n$*, if the greedy matching sends a pizza from* x_i *to* y_j*, place a circle with radius* $|x_i - y_j|/2$ *at* $(x_i + y_j)/2$*. These* n *circles do not touch.*

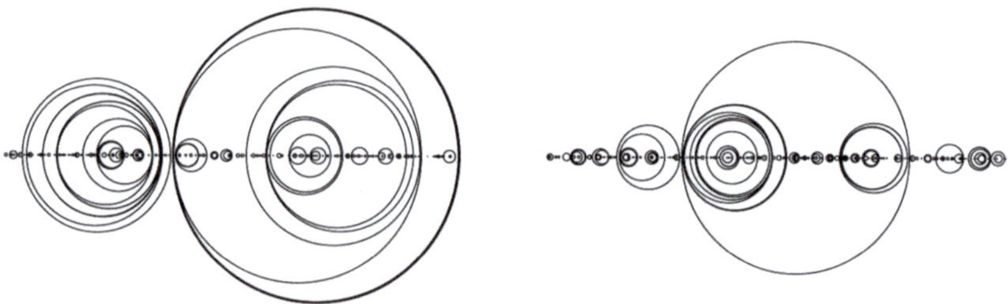

Fig. 17.4: Two examples of 500 houses and 500 pizzerias: the greedy matching produces the location of 500 circles (many of which are small) that do not touch.

This result is actually not tremendously difficult to prove, and we leave it as an exercise. It is perhaps good to see two examples of what these circles can look like, so we refer to Fig. 17.4. For more information, have a look at recent work with Andrea Ottolini, who delivered pizzas in Italy when he was a teenager [218].

17.2 Matching residents: the Gale–Shapley theorem

When matching pizzerias with houses, the way we measured the *quality* of a matching was obvious: total amounts of miles driven. This is an example of a quantitative metric, things are measured in terms of numbers. The purpose of this section is to describe a different type of matching problem. Suppose we have n high school students who are about to apply for college. Let us make this concrete and suppose we have $n = 3$ students, Ann, Mark and Susan, who apply to three schools: Hogwarts, the Starfleet Academy, and the Jedi Temple. We use $A \succ B$ to denote that A is preferred over B. Ann always wanted to be a witch and does not care for lightsabers, Mark and Susan have their own preferences

$$
\begin{array}{ll}
\textbf{Ann:} & \text{Hogwarts} \succ \text{Starfleet} \succ \text{Jedi Temple} \\
\textbf{Mark:} & \text{Hogwarts} \succ \text{Jedi Temple} \succ \text{Starfleet} \\
\textbf{Susan:} & \text{Starfleet} \succ \text{Jedi Temple} \succ \text{Hogwarts}
\end{array}
$$

The three Institutions, in turn, have their own individual ranking of the applicants in terms of suitability, and maybe their ranking looks like this

$$
\begin{array}{ll}
\textbf{Hogwarts:} & \text{Ann} \succ \text{Susan} \succ \text{Mark} \\
\textbf{Starfleet:} & \text{Mark} \succ \text{Ann} \succ \text{Susan} \\
\textbf{Jedi Temple:} & \text{Ann} \succ \text{Susan} \succ \text{Mark}
\end{array}
$$

What would be a good way to decide which school each student ends up attending? David Gale and Lloyd Shapley [95] point out that while many assignments make sense, there are some that are problematic: suppose a student β really wants to attend college A and college A really wants β as a student and we somehow send β to another college and send a different student α to A.

> Suppose the situation described above did occur. Applicant β could indicate to college A that he would like to transfer to it, and A could respond by admitting β, letting α go to remain within its quota. Both A and β would consider the change an improvement. The original assignment is therefore "unstable" in the sense that it can be upset by a college and applicant acting together in a manner which benefits both. (Gale–Shapley, [95])

We can illustrate this in our setting. A problematic assignment would be

$$
\text{Ann} \leftrightarrow \text{Starfleet} \qquad \text{Mark} \leftrightarrow \text{Jedi Temple} \qquad \text{Susan} \leftrightarrow \text{Hogwarts}
$$

because, in this assignment, the following scenario becomes possible: Ann could approach Hogwarts and point out that she would really like to go to Hogwarts (she prefers it over the Starfleet Academy), Hogwarts would then realize that they would prefer to have Ann has a student (who they rank higher than Susan). Both Ann and Hogwarts realize that they would be better off if they just ignored the centralized assignment completely. A matching between students and schools is *unstable* if a student and a school could improve their situation by ignoring the system. Our goal is to find stable matchings. This is relevant once a year in the United States when students finishing medical school are applying to be residents in hospitals: the students submit their preference ordering and, conversely, residence programs rank applicants. All of this is organized by the *National Resident Matching Program*, also known as *The Match*, and the results are released on *Match Day*.

A formal description. Here is a formal description. We are given

1. n students who are applying to n different schools.
2. Each of the n students has an (individual) ranking of the n schools in terms of how much they would like to go there,
3. while the n schools also have a ranking of the n applicants.

We are interested in finding a matching between students and schools that is *stable*. A fundamental fact is that such a stable matching always exists – independently of how students order the schools or how schools order the students.

Theorem (Gale–Shapley [95]). *A stable matching exists.*

This result is so very nice that it's worth a Nobel Prize, awarded to Lloyd Shapley in 2012 *for the theory of stable allocations and the practice of market design*, specifically it is pointed out that *Shapley and his colleagues derived specific methods – in particular, the so-called Gale–Shapley algorithm – that always ensures a stable matching*.

Proof of the Gale–Shapley theorem. This proof is completely constructive. We explicitly describe how to find a stable matching by playing a game. There are several rounds, and each round the following happens:

1. Schools with an open slot, meaning schools that do not currently have a student assigned to them, make an offer to the candidate they prefer among those they have not yet made an offer to.
2. Any candidate who receives one of these offers does the following: if they are not yet assigned to a school, they (provisionally) accept the offer. If they are already assigned to a school and prefer the new offer over the one they have, they match to the new school (leaving their previous school with an open position). If they are already assigned to a school and do not prefer the new offer to their current program, they stick with their old school.
3. This is repeated until everybody is matched.

We have to prove two things: that every student gets matched with a school and that the matches that come out of this procedure end up being stable.

1. Everybody gets matched. There are n students and n programs. Once a student gets into a program, they remain matched to some program for all time (they may 'upgrade' schools along the way but they will never ever again be not matched).

A school that is currently unmatched, that has no student assigned to it, always makes an offer. Since the number of students is the same as the number of schools, everybody eventually receives at least one offer.

2. The matches are stable. Suppose (A, B) are an unstable pair. That means that the school B is matched to a student A_2, who they rank lower than A. So they would prefer having A as their student. That means that B must have made an offer to A before (because schools make offers in decreasing order of ranking). The only way that student A would not have accepted the offer from school B is if they were already matched to a school that they rank higher than A (otherwise they would have switched to A) – but then (A, B) cannot be an unstable pair. $\qquad\square$

Consider the above example of Ann, Mark and Susan. Hogwarts offers Ann a spot. Ann has no current school and (provisionally) accepts Hogwarts. The Starfleet Academy offers Mark a spot. Mark has no current school and accepts. The Jedi Temple offers Ann a spot. Ann already committed to Hogwarts and prefers Hogwarts over the Jedi Temple and declines. The Jedi Temple then offers Susan a spot. Susan had not previously been offered a spot and accepts. We arrive at

$$\text{Ann} \leftrightarrow \text{Hogwarts} \qquad \text{Mark} \leftrightarrow \text{Starfleet} \qquad \text{Susan} \leftrightarrow \text{Jedi Temple}.$$

We observe that we could also turn the reasoning around: instead of schools offering spots to students, students could offer their attendance to schools, the entire setup is perfectly symmetric. In that case, Ann would offer to attend Hogwarts, and Hogwarts, not yet having a student, would accept. Mark would offer to attend Hogwarts, but Hogwarts prefers Ann over Mark and declines. Susan offers to attend the Starfleet Academy, and the Starfleet Academy accepts. Mark offers to attend the Jedi Temple, and the Jedi Temple accepts. We arrive at

$$\text{Ann} \leftrightarrow \text{Hogwarts} \qquad \text{Mark} \leftrightarrow \text{Jedi Temple} \qquad \text{Susan} \leftrightarrow \text{Starfleet}.$$

A more careful inspection shows that in the first case, schools offering spots to students, the schools end up happy. In the second case, students offering their attendance to schools, the students end up happier.

17.3 Exercises

1. Learn about the Borda count. Introduced by Jean-Charles de Borda in the paper just before Monge's paper, it was first invented by Nicholas of Cusa (Cusanus) more than 300 years earlier. Cusanus, in his 1446 *Coniectura de ultimis diebus* ('Conjecture of the Last Days') [65] gives a fairly precise range for the Apocalypse (*And this will be after the year 1700, before the year 1734*). Cusanus did note that there are a lot of different predictions for when the apocalypse will happen (*Many other things have been written about this matter [...] each one abounds in his own interpretation, agreeing with no one else*), however, he ultimately remained

convinced that he is right, *I have diligently investigated their writings, and I have found nothing in them like what I have just set down* [191].

2. Consider pizzerias in $\{0, 0.3, 0.7, 1\}$ and $\{0.1, 0.65, 0.9, 1\}$. What is the minimal transport cost when sending a pizza from x to y costs $|x - y|$?

3. The same question as (2) except the transport cost is now $|x - y|^{1/10}$.

4. Yet another way of assigning pizzas and houses is the Dyck matching (see [44] or [218] for a short explanation). Learn about the Dyck matching and compute it for the example above.

5. Prove that the n circles created from the greedy matching do not cross.

6. Ponder how the optimal transport problem and the matching problem are tied together. One could think of each pizzeria having a preference ranking of houses (the house that is the closest has the highest ranking, the second-closest the second-highest and so on).

Chapter 18
Conflict and Cooperation

His perfect skill the wondering gazers eyed,
The game as yet unseen, as yet untried
(Homer, *The Odyssey* [141])

18.1 Early History of Game Theory

The early history of Game Theory depends a little bit on the definition of game. As discussed in §16, probability theory started by asking questions about games, we recall the title of the 1657 book by Huygens *De Ratiociniis in Ludo Aleae* (On Reasoning in Games of Chance). Many of the subsequent studies in probability theory were motivated by the setting of games but usually asymmetrically: gambling against the all-powerful casino. Game Theory, as the term is now understood, tries to understand what happens when two players meet and compete in the setting of a game. What are rational strategies? How should we play?

> In all of man's written record there has been a preoccupation with conflict of interest; possibly only the topics of God, love, and inner struggle have received comparable attention. (D. Luce and H. Raiffa's 1957 *Games and Decisions*, [180])

Questions of this type arose early: Cournot [62], in 1838, considered two competing companies and how they would react to each other. In 1912, Zermelo published *On an Application of Set Theory to the Theory of the Game of Chess* [310] and notes that, in a game of a chess there is an *optimal strategy*. This means that either the player moving first (white) can execute a series of moves such that they can either force a win or a draw or, conversely, that the player moving second (black) has such a series of moves – we simply do not know who does. We also do not know what the strategy would look like. It is widely believed that the player moving first has an advantage and can force either a win or at least a draw; this is substantiated by statistics and quotes such as the one ascribed to chess grandmaster Efim Bogoljubow: *When I am White I win because I am White. When I am Black I win because I am Bogoljubow.* The most important contribution of Zermelo is perhaps the idea of a *strategy*, a plan of action that takes into account all possible actions of the other player.

S. Steinerberger, *The Unreasonable Elegance of Mathematics*,
Springer Undergraduate Mathematics Series, https://doi.org/10.1007/978-3-032-03815-9_18

18.2 Two Thirds of an Average

Why did it take until the 20th century for modern Game Theory to begin? Games of Chance were already considered in the 1660s. The long delay may have been because of a conceptual difficulty: imagine playing a game against someone else. If I play a certain strategy and my opponent knows that I am playing a certain strategy, they can counter with a different strategy – but I know that they know that they are countering my strategy, which suggests that I should play a strategy countering their counterstrategy, and things become messy.

> **Guess** 2/3 **of the average.** This game can be played with any number of players. Each player thinks of a number between 0 and 100 and writes it on a piece of paper. The pieces of paper are collected, the player whose number is closest to 2/3 of the average of the numbers wins.

Imagine you are playing this game. The first realization is that if everybody picks 100, the average would be 100 and 2/3 of the average would be 66.6. It makes no sense to pick a number that is larger than 66.6. The optimal choice of a number is therefore somewhere between 0 and 66.6. However, if everybody else also realizes this, then nobody is going to pick a number larger than 66.6. In that case, the average is at most 66.6 and 2/3 of the average cannot exceed 44.4. However, if everyone else realizes this, well, then nobody is going to pick a number larger than 44.4 and the story continues all the way down to zero.

Fig. 18.1: The game in Dinosaur Comics (no. 1360, © Ryan North).

This illustrates one of the dangers of trying to get wrapped up in thinking about thinking about thinking about thinking. In practice, one could repeat this chain of reasoning an arbitrary number of times and end up concluding that the only reasonable thing is to write the number 0. However, that's not what people do [205]. In fact, the author tried it out with a large number of people and two-thirds of an average ended up being 31.

18.3 Prisoner's Dilemma and Global Warming

> [...] that man to man is an arrant wolfe.
> (Thomas Hobbes, *De Cive* [137])

> the proverb contains two major flaws. First, it fails to do justice to canids, which
> are among the most gregarious and cooperative animals on the planet [...]
> (Frans de Waal, *Primes and Philosophers* [291])

One of the classic examples of Game Theory is the Prisoner's Dilemma. It may seem like a cute story, but it encapsulates a fundamental problem. Recall that the term *Faustian bargain* refers to Faust's legendary deal with the Devil (*When thus I hail the Moment flying // "Ah, still delay – thou art so fair!" // Then bind me in thy bonds undying,// My final ruin then declare!* [112]). With this in mind, a nice description of the Prisoner's Dilemma is as follows

> Two members of a criminal gang are arrested and imprisoned. Each prisoner is in solitary confinement with no means of speaking to or exchanging messages with the other. The police admit they don't have enough evidence to convict the pair on the principal charge. They plan to sentence both to a year in prison on a lesser charge. Simultaneously, the police offer each prisoner a Faustian bargain. If he testifies against his partner, he will go free while the partner will get three years in prison on the main charge. Oh, yes, there is a catch. If both prisoners testify against each other, both will be sentenced to two years in jail. [236]

		B's Choice	
		Cooperate (C)	Defect (D)
A's Choice	Cooperate (C)	(2, 2)	(0, 3)
	Defect (D)	(3, 0)	(1, 1)

Table 18.1: The Prisoner's Dilemma payoff matrix. Each player has to either cooperate with the police or defect. The result (a, b) means that A is spending a years in prison and B is spending b years in prison.

One analyzes this by listing all four possible options. Each of the two prisoners, A and B, can choose to either cooperate with the police (C) or to defect (D). Both players are trying to minimize the number of years they spend in jail. The problem can be neatly summarized as follows.

1. A is always better off by cooperating, independently of what B is doing.
2. B is always better off by cooperating, independently of what A is doing.
3. If they both cooperate with the police, they end up in a suboptimal solution, 2 years in jail for both, compared to what they could have achieved (1 year each).

The Prisoner's dilemma illustrates a setting where individuals pursuing their own self-interest leads to a worse outcome for everyone. It may appear not to be relevant to our daily life unless you have plans to enter the criminal sector. So let us change the formulation! Two countries, A and B, have the choice to either reduce carbon emissions (Reduce) or not to reduce said emissions (Not Reduce). Both countries have an interest in solving the problem of global warming. However, reducing carbon emissions has an economic impact, and, from a (short-term) economic point of view, they would prefer not to do it. If A decides not to reduce carbon emissions and B does, well, that is great news for A. A will benefit from the reduced carbon emissions of B while not feeling any economic impact themselves. We see that the structure of the problem can again be described by the Prisoner's Dilemma (see Table 18.2): independently of whether country B reduces carbon emissions, country A (individually) is always better off when not reducing carbon emissions.

		B's Choice	
		Reduce	Not Reduce
A's Choice	Reduce	(2, 2)	(0, 3)
	Not Reduce	(3, 0)	(1, 1)

Table 18.2: The carbon emission problem.

The best possible economic situation for each individual country would be if all the other countries were to reduce carbon emissions while they themselves alone could reap the benefits of ignoring the problem of global warming completely. However, if every country pursues their own self-interest, then the overall outcome will be, mildly put, suboptimal in the long run. An abstract description of the mechanism at play was given by William F. Lloyd [175], discussing the perils of collaboration

> Suppose the case of two persons agreeing to labour jointly, and that the result of their labour is to be common property. Then, were either of them, at any time, to increase his exertions beyond their previous amount, only half of the resulting benefit would fall to his share; were he to relax them, he would bear only half the loss. If, therefore, we may estimate the motives for exertion by the magnitude of the personal consequences expected by each individual, these motives would in this case have only half the force, which they would have, were each labouring separately for his own individual benefit. Similarly, in the case of three partners, they would have only one third of the force—in the case of four, only one fourth—and in a multitude, no force whatever. (Lloyd, *Two Lectures on the Checks to Population* [175])

Once one knows about the problem, one starts seeing it everywhere! The author was once teaching a class in which students expressed an interest in seeing *all* possible solutions to the homework problem. Good problems can often be solved in many ways, it is a reasonable request. The author pointed that out that since all thirty students had already done the homework, the easiest would be to simply share all their different solutions and provided an online platform where this could be done.

1. Each individual student is only interested in the homework uploaded by other people. They already know their own solution.
2. Since it takes work to make one's own work available online and it is independent of whether other students will do so, each individual student would prefer not to upload their own work.

The result was predictable: not a single student uploaded their solution, and they all had to listen to a surprise lecture about the Prisoner's Dilemma.

18.4 Robert Axelrod's Tit for Tat

So how do we escape the trap set by the Prisoner's Dilemma? One way out may have been suggested by the political scientist Robert Axelrod (1943–) in his 1984 book *The Evolution of Cooperation* [14]. Axelrod hosted a tournament where strategies would repeatedly play the Prisoner's Dilemma against each other. Each strategy would accumulate a number of points depending on well they did against other strategies. So what type of strategies are there? The two simplest strategies are

- Always cooperate.
- Always defect.

If two always-cooperate strategies meet, they both cooperate and do extremely well. If an always-cooperate meets an always-defect, one ends up exploiting the other. If two always-defect strategies meet, they try to mutually exploit each other and both perform very poorly. Many complicated strategies have been proposed. Some try to first understand how their opponent thinks and then try to exploit that. Consider, for example, the strategy DOWNING (named after [71]):

> This rule, called DOWNING, is a particularly interesting rule in its own right. It is well worth studying as an example of a decision rule which is based upon a quite sophisticated idea. [...] it is based on a deliberate attempt to understand the other player and then to make the choice that will yield the best long-term score based upon this understanding. [...] To judge the other's responsiveness, DOWNING estimates the probability that the other player cooperates after it (DOWNING) cooperates, and also the probability that the other player cooperates after DOWNING defects. For each move, it updates its estimate of these two conditional probabilities and then selects the choice which will maximize its own long-term payoff under the assumption that it has correctly modeled the other player. (Axelrod [14])

This is a very complicated strategy *based on an "outcome maximization" principle originally developed as a possible interpretation of what human subjects do in the Prisoner's Dilemma laboratory experiments.* The biggest surprise of the tournament was the winner: a strategy known as Tit-for-Tat submitted by Anatol Rapaport. Tit-for-Tat starts by cooperating and then repeats the strategy played by the opponent in the previous round. As long as the opponent cooperates, so does Tit-for-Tat. Simultaneously, it cannot be exploited: if the opponent defects, so will Tit-for-Tat. As noted by Axelrod, this is part of a larger pattern. *The tournament results from chapter 2 are very striking. The single best predictor of how well a rule performed was whether or not it was nice, which is to say, whether or not it would ever be the first to defect* [14]. It pays off to be nice, it pays off to give the benefit of the doubt (never be the first to defect) and it pays off to forgive (independently of what happened in the past, if a strategy cooperates, Tit-for-Tat will cooperate in the next round). This may even explain why we see cooperation in nature: it pays off!

18.5 Conflict: von Neumann and Nash

Modern Game Theory starts with John von Neumann's 1928 paper *Zur Theorie der Gesellschaftsspiele* (On the Theory of Games of Strategy) [211]. John von Neumann (1903–1957) is not nearly as well known as he deserves to be, *his name should be as well known as Albert Einstein's* [164]. Born and raised in Budapest, Hungary, he was a singular phenomenon early on; one of his best friends from childhood, Eugene Wigner, later became a famous physicist. *In a 1963 interview [...], the year he won his Nobel Prize, Wigner seemed almost to have an inferiority complex toward the dead Johnny. "You have a good memory?" asked Kuhn. "Not like von Neumann's," replied Wigner. Kuhn was soon saying that "it must have been a shattering experience to have grown up with von Neumann however bright one is"* [184]. His achievements are too many to list. Starting out in set theory and logic, he quickly moved on to prove the Ergodic Theorem and did some foundational work in Measure Theory. These *contributions amount roughly to one tenth of von Neumann's scientific publications. As to their quality, it seems to be safe to say that if von Neumann had never done anything else, they would have been sufficient to guarantee him mathematical immortality* [123]. Besides fundamental contributions in physics, including his 1932 book *Mathematical Foundations of Quantum Mechanics* [212], he started modern Game Theory and Mathematical Economics, first with his 1928 paper *On the Theory of Games of Strategy* [211] followed by his 1944 book *Theory of Games and Economic Behavior* [213] written with Oskar Morgenstern. Von Neumann was a foundational figure in the modern theory of Computers, produced some of the first algorithms (merge sort), is sometimes credited with starting the field of cellular automata (which we saw many chapters ago), and *for scientific computing he may be the most influential researcher of all time* [121]. Claude Shannon, the inventor of Information Theory, called him *the smartest person I've ever met* [266], it has been argued that *No other mathematician in this century has had as deep and lasting an influence on the course of civilization* [246]. Returning to Game Theory, von Neumann [211] showed the *von Neumann Minimax Theorem*: every zero-sum game with a finite number of strategies that is played by two adversarial players has a value, an outcome that can always be enforced by one player and not surpassed by the other. The second big result in non-cooperative game theory is the notion of the *Nash equilibrium* developed by John Nash (1928–2015) in his 1950 dissertation. Both of these notions, the Minimax Theorem and the Nash Equilibrium, describe theories of conflict. It is somewhat poetic that von Neumann and Nash met and that this meeting was not without conflict.

> He [John von Neumann] listened carefully, with his head cocked slightly to one side and his fingers tapping. Nash started to describe the proof he had in mind for an equilibrium in games of more than two players. But before he had gotten out more than a few disjointed sentences, von Neumann interrupted, jumped ahead to the yet unstated conclusion of Nash's argument, and said abruptly, "That's trivial, you know. That's just a fixed point theorem."
> (Sylvia Nasar, [208])

After receiving the Nobel Prize in Economics, Nash recalls that his *game theory ideas [...] deviated somewhat from the "line" (as if of "political party lines") of von Neumann and Morgenstern's book* and also wrote *I was playing a non-cooperative game in relation to von Neumann rather than simply seeking to join his coalition. And of course, it was psychologically natural for him not to be entirely pleased by a rival theoretical approach* [208].

18.6 Calculating Fairness: Utilitarianism

Game Theory has a philosophical precursor: Utilitarianism. A remarkable philosopher in this field was John Stuart Mill (1806–1873) whose father raised him with the somewhat unusual goal that he may turn out to be a genius. Mill was subjected to the most unusual education

> I have no remembrance of the time when I began to learn Greek; I have been told that it was when I was three years old. [...] I learnt no Latin until my eighth year. At that time I had read, under my father's tuition, a number of Greek prose authors, among whom I remember the whole of Herodotus, and [...] some of the lives of the philosophers by Diogenes Laertius [...] The only thing besides Greek, that I learnt as a lesson in this part of my childhood, was arithmetic: this also my father taught me: it was the task of the evenings, and I well remember its disagreeableness. (John Stuart Mill, Autobiography [196])

Mill published *Utilitarianism* [197], which starts by lamenting *the little progress which has been made in the decision of the controversy respecting the criterion of right and wrong* and argues that things should not be this complicated. *The creed which accepts as the foundation of morals, Utility, or the Greatest Happiness Principle, holds that actions are right in proportion as they tend to promote happiness, wrong as they tend to produce the reverse of happiness* [197]. The idea is relatively simple: if we have to choose among various options, we should choose the one that maximizes the total amount of *happiness* (or *utility*) in the world. A concrete example would be as follows: suppose we have one cake and three hungry people, Ann, Bob and Charlie. How should we divide the cake? The first question is perhaps: how happy is a person when they receive x units of cake? The answer probably looks something like Fig. 16.3: twice as much cake does not make one twice as happy! So let us suppose the happiness caused by x units of cake is described by $h(x) = \sqrt{x}$. We then have the following general result about the total amount of happiness.

Theorem. *If $0 \leq x_1, \ldots, x_n \leq 1$ is the division of $x_1 + x_2 + \cdots + x_n = 1$ unit of cake, then the total happiness satisfies*

$$\sqrt{x_1} + \sqrt{x_2} + \cdots + \sqrt{x_n} \leq \sqrt{n}$$

with equality if and only if $x_i = 1/n$ for all $1 \leq i \leq n$.

This is a nice result: the total amount of happiness is maximized when we are as fair as possible and give everyone the same amount of cake! The proof of the Theorem can be achieved by a nice trick (the Cauchy–Schwarz inequality), but we will not go into details. The utilitarian approach can also be used to solve more complicated asymmetric questions: suppose Ann and Bob have to share a cake. Ann is not as hungry as Bob. Her happiness with x units of cake is $h_1(x) = x^{2/3}$ while Bob is as happy as in the example above, $h_2(x) = \sqrt{x}$. We note that for $0 \leq x \leq 1$ we have $h_1(x) = x^{2/3} < \sqrt{x} = h_2(x)$, so the same amount of cake makes Bob happier than it does Ann. If we give x units of cake to Ann and the rest of the cake to Bob, the total amount of happiness is $x^{2/3} + \sqrt{1-x}$. John Stuart Mill argues that we should choose $0 < x < 1$ so as to maximize this function, and some work shows that the 'optimal' solution is

$$x = \frac{3853 - \dfrac{214623}{\sqrt[3]{262144\sqrt{997} - 7412963}} + 9\sqrt[3]{262144\sqrt{997} - 7412963}}{4096} \sim 0.59989\ldots$$

Ann would receive around 60% of the cake, Bob the remaining 40%. Ann's happiness would be ~ 0.711 and Bob's happiness would be ~ 0.632. Does this seem fair? It certainly maximizes the *sum* of the happiness in the world.

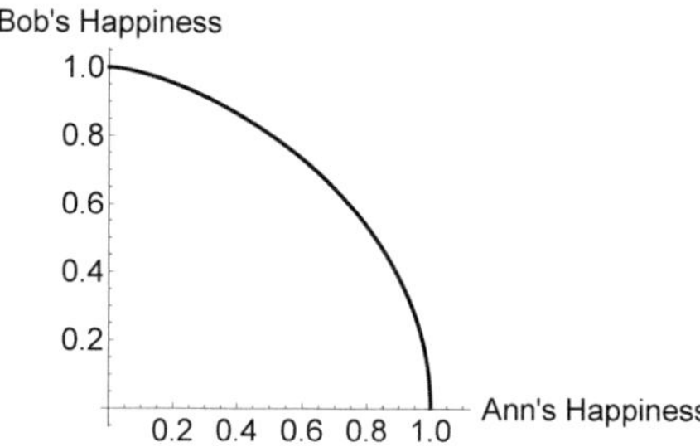

Fig. 18.2: Ann and Bob have to divide happiness. They have to choose a point (x, y) on the curve, Ann gets x happiness and Bob gets y happiness.

18.7 The Bargaining Problem

Suppose we have two players, Ann and Bob, and they are facing a decision how to divide a certain good (money, cake, vacation time, ...). However, Ann and Bob are different people and experience different levels of happiness. A good way to visualize the dilemma is to keep track of how happy each choice would make the two players. If we give x cake to Ann and $1 - x$ cake to Bob, their respective happiness is $(x^{2/3}, (1 - x)^{1/2})$. Substituting $y = x^{2/3}$, we note that $x = y^{3/2}$, and we can describe the curve also as $(y, (1 - y^{3/2})^{1/2})$ for $0 \leq y \leq 1$ which corresponds to the graph of the function $f(x) = \sqrt{1 - x^{3/2}}$ for $0 \leq x \leq 1$ shown in Fig. 18.2. The big problem is now: Ann and Bob (or some external party) has to decide what point on the happiness curve to choose. We first note that, technically, one could also choose other points: Ann and Bob need not split the entire cake, they could each take $1/3$ of the cake and throw the remaining $1/3$ in the trash bin. However, since each of their respective happiness levels is increasing with the amount of cake they receive, they will never throw anything away. This No-Trash-Can principle is also known as *Pareto efficiency*, after Vilfredo Pareto (1848–1923). In practice, the only reasonable choice is going to be on the curve shown in Fig. 18.2. But which point(s) are fair? This is

The Fundamental Problem of Fairness. Which division should we choose?

Ann and Bob have opposing interests: whatever we end up choosing, Ann will always ask for a point further to the right. Bob will always want a point that is further up. Being restricted to being on the curve, we cannot make both of them happy at the same time. So what would be *fair*? Well, fairness is in the eye of the beholder: it depends on who you ask. We start with two possible solutions.

1. **Egalitarian solution.** We do not care about the structure of the game: both players should receive the exact same amount of happiness.
2. **Utilitarian solution.** We try to maximize the sum of happiness in the world.

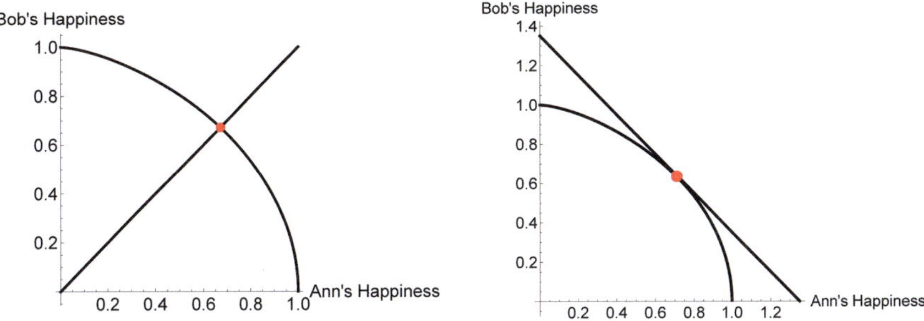

Fig. 18.3: Left: the egalitarian approach, both receive ~ 0.671 happiness. Right: the utilitarian approach: Ann receives ~ 0.711 and Bob receives ~ 0.632 (as above).

The egalitarian approach amounts to intersecting the happiness curve with the line $y = x$ while the utilitarian approach means, geometrically, that we try to find the largest value c such that $y = c - x$ still intersects the line (in that case the total happiness in the world is $x + y = c$), see Fig. 18.3. These two solutions are different: which one is *better*? It depends on who you ask. Some people would argue that everyone has the same unalienable right to happiness, they would prefer the egalitarian solution. Others would argue that the game favors Ann and *when equality is given to unequal things, the resultant will be unequal* (Plato [232]). More precisely, if we consider the egalitarian solution (x^*, x^*) with $x^* \sim 0.671$, then we see that the slope of the curve in the point is, locally, roughly -0.9. This means that Bob could sacrifice 0.009 of his happiness and Ann would receive 0.01 of happiness in return. Ann did not ask for the game to be created this way, why should she sacrifice a disproportional amount of happiness?

People have discussed this for a *long* time. The Ann-Bob problem is useful: many people try to sell us a single solution for all the problems in the world. If we all just follow philosophy X or political idea Y, then we will all live in harmony! But the fundamental problem is right there: how much happiness do we give Ann, and how much happiness do we give Bob? No matter how beautiful our ideas may be, here is a concrete problem in demand of a concrete solution, and it is not going to go away! Consider an idea of John Rawls (1921–2002): the *veil of ignorance*. Suppose we want to discuss how to organize a society; a familiar problem is that farmers will argue on behalf of the farmers, and smiths will argue on behalf of smiths. Enter the philosopher John Rawls.

> The idea of the original position is to set up a fair procedure so that any principles agreed to will be just. [...] Somehow we must nullify the effects of specific contingencies which put men at odds and tempt them to exploit social and natural circumstances to their own advantage. Now in order to do this I assume that the parties are situated behind a veil of ignorance. They do not know how the various alternatives will affect their own particular case and they are obliged to evaluate principles solely on the basis of general considerations. (John Rawls, *A Theory of Justice* [240])

Imagine, Rawls says, the task of designing a society in which you will occupy a place that is not clear to you. *It is assumed, then, that the parties do not know certain kinds of particular facts. First of all, no one knows his place in society, his class position or social status; nor does he know his fortune in the distribution of natural assets and abilities, his intelligence and strength, and the like. Nor, again, does anyone know his conception of the good, the particulars of his rational plan of life, or even the special features of his psychology such as his aversion to risk or liability to optimism or pessimism* (Rawls [240]). How would we design such a society? As happens frequently, the idea predates Rawls. John Harsanyi, who ended up receiving a Nobel Prize in Economics, writes in 1953 that *If somebody prefers an income distribution more favorable to the poor for the sole reason that he is poor himself, this can hardly be considered as a genuine value judgment on social welfare* [127] and adds

> Now, a value judgment on the distribution of income would show the required impersonality to the highest degree if the person who made this judgment had to choose a particular income distribution in complete ignorance of what his own relative position (and the position of those near to his heart) would be within the system chosen. (Harsanyi [127], 1953)

It is a great idea but not easy to apply. Imagine you are going to wake up and find yourself to be Ann or Bob (with equal likelihood). What solution would you prefer? The egalitarian solution guarantees you 0.671 happiness. The utilitarian solution guarantees you either 0.632 or 0.711 happiness, with equal likelihood, and thus an *expected* happiness of $(0.632 + 0.711)/2 = 0.6715$, which is slightly larger than the egalitarian solution. This sounds familiar, we are back in the realm of probability theory, expected value and risk aversion. If you end up being Ann/Bob many times over, you would presumably choose the utilitarian solution, it has the highest expected value. This raises a philosophical question: if one believes in reincarnation, does this force one to be a Utilitarian? It is an interesting question [263].

18.8 The Nash Solution

John Nash [209] not only studied the problem but proposed a nontrivial change of perspective that is (almost) completely devoid of philosophy. Let us, he proposes, study the underlying *axioms*.

> We shall develop the theory by giving conditions which should hold for the relationship between this solution point and the set, and from these deduce a simple condition determining the solution point. (Nash [209], 1950)

Instead of trying to argue for or against a particular solution, we may ask, more abstractly, *what properties do we want the solution to satisfy*? And once we finalize what these desired properties are, what solutions exhibit all these properties? This changes the landscape of the problem: instead of arguing loudly in favor of the egalitarian solution or praising the fairness of the utilitarian solution, we may instead ask: what properties do we want a solution to have? Nash proposed four key properties

1. (Invariance under Affine Transformations.) If we change our measurement for Ann from 'happiness' to 'micro-happiness' (one happiness is 1000 micro-happiness), then this changes the scale and the numbers but the solution should not change.
2. (Pareto Optimality.) The solution has to be Pareto optimal (we do not throw any cake in the trash, we pick a point on the curve).
3. (Independence of Irrelevant Alternatives.) If a solution declares a point to be fair and we change the possible happiness outcome somewhere else, the solution should not change: if we agree to split the cake 50-50 and someone comes along and forbids us to split the cake 70-30, well, that is fine, we were going to split it 50-50 anyway, so that does not matter.
4. (Symmetry.) If we call Bob 'Ann' and Ann 'Bob', everything flips.

Nash proved that the only solution that satisfies all these properties is to maximize *the product of the happiness of Ann and Bob*. This corresponds to finding the rectangle with maximal area whose diagonal points are in $(0, 0)$ and somewhere on the happiness curve. Here, Ann receives ~ 0.688 and Bob receives ~ 0.654.

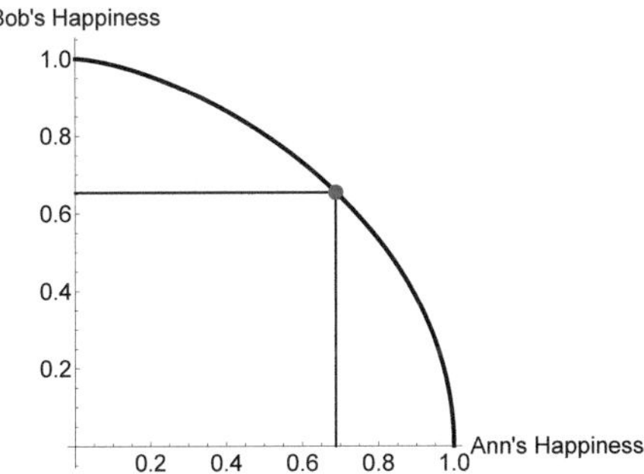

Fig. 18.4: The Nash Solution: find the rectangle with the largest area. In this case, the Nash solution awards Ann ~ 0.688 happiness, Bob receives ~ 0.654 happiness.

Whether we believe the Nash solution to be fair depends on whether we believe the four properties to be fair. Property (1) makes sense. Suppose we divide a sum of money, and then Ann decides to receive her money not in Dollars but in Euros; surely that should not mean that she will receive more or less money, she should receive exactly the same Dollar amount transferred into Euros. Pareto Optimality, Property (2), is uncontroversial. Symmetry, Property (4), is a natural embodiment of fairness. The Nash solution has not gone unchallenged. The main objection is usually towards Property (3): if the game chooses a solution and we change the outcome of the game somewhere else, the solution should not change. However, one could argue, if you change the rules of the game, you change the entire game. If the game changes, then the relative bargaining positions of the players also change – in fact, you could argue, everything changes. And if you change the game, surely it should be within the realm of the possible to reevaluate the solutions?

18.9 Exercises

1. Find some *very* patient friends and actually play 'Two Thirds of an Average', and see what numbers people end up choosing.
2. Find more instances of the Prisoner's Dilemma. It can be found all over history. Plautus in his *Asinaria* (A Comedy of Asses) has the Trader say

   > **Trader.** But notwithstanding, never will you induce me today to trust this money to you, a stranger. "Man is no man, but a wolf, to a stranger." (Plautus [233])

   Another example is Machiavelli discussing politics in 1532

   > It is unquestionably very praiseworthy in princes to be faithful to their engagements [...] I could give numerous proofs of it, and shew how many engagements and treaties have been broken by the infidelity of princes; the most fortunate of whom has always been he who best understood how to assume the character of the fox. The object is to act his part well, and to know how in due time to feign and dissemble; and men are so simple and so weak, that he who wishes to deceive easily finds dupes. (Machiavelli, *The Prince* [183])

3. Prove that if happiness is given by $\sqrt{x}$ and we have to divide a cake of fixed size among two people, then the utilitarian solution is to split the cake evenly.
4. Two people, Ann and Bob, divide one unit of cake. Ann gets x, creating happiness x and Bob gets $1 - x$, creating happiness $\sqrt{1 - x}$. Determine the egalitarian, the utilitarian and the Nash solution.
5. Argue both in favor of and against Nash's Property 3, the Independence of Irrelevant Alternatives. Which of the two arguments do you find more convincing? What would Buddha, Karl Marx and Ludwig von Mises say? Ethical and economic theories are not necessarily easy to translate into a setting where concrete numbers are required.
6. The next time someone convinces you that abstract principle Z will solve all the problems in the world, ask them to explain how abstract principle Z would divide the cake between Ann and Bob.
7. Find out about the Kalai–Smorodinsky solution [148]. How does it compare to Nash? What are the arguments in favor of it, what are the arguments against it?

Chapter 19
The End of Knowledge

19.1 Three Humiliations

In 1917, Sigmund Freud published a short little paper, *A Difficulty for Psychoanalysis* [92], in which he points out that Psychoanalysis may face substantial opposition for a particularly interesting reason: it is humiliating.

A Difficulty for Psychoanalysis
by Sigm. Freud (Vienna)

> I will, already at the very beginning, clarify that I do not mean an intellectual difficulty, something that would make Psychoanalysis difficult to grasp for the recipient (reader or listener), but an affective difficulty: something, through which Psychoanalysis is alienating, so that he, the recipient, will be less likely to be interested or inclined to believe in it. Both difficulties have the same result. If sympathy is lacking, understanding will not come easily. For the sake of the reader, I must paint some broader strokes.... (Sigmund Freud [92])

Freud argues that Psychoanalysis, as an idea, will face resistance because it goes against a fundamental aspect of what we truly believe: we are at the center of the universe, we are important and we know who we are. All three have been disproven over the years. Freud, with a certain degree of joy, distinguishes between the cosmological, the biological, and the psychological insult.

1. Man first believed that his home, the earth, was at rest in the center of the universe, while the Sun, Moon and planets moved in circular orbits around the earth. [...] The destruction of this narcissistic illusion is linked to the name and work of Nicolaus Copernicus in the sixteenth century. [...] when it first gained acceptance, human self-love experiences its first, cosmological, insult (Freud, [92]).

2. Man, in the course of cultural development, assumed the role of master over the animals. [...] We all know that the research of Charles Darwin [...] little more than half a century ago, put an end to this arrogance of man. [...] This is the second, biological, insult of human narcissism [...] (Freud, [92])

3. The third insult, which is of a psychological nature, is probably the most serious [...] Man, even though humiliated outside, feels sovereign in his own soul. [...] However, the Self is not master in its own house. [...] (Freud, [92]).

Freud argues that this will cause a big difficulty for Psychoanalysis, *if sympathy is lacking, understanding will not come easily*. A nice variation, attributed to Upton Sinclair, is *It is difficult to get a man to understand something, when his salary depends on his not understanding it*. Freud, ever the diligent scholar, points out that none of these ideas are new: already Aristarchus of Samos pointed out, more than 2000 years ago, that the earth may be much smaller than the sun and revolve around it. Freud mentions Charles Darwin together with *his coworkers and predecessors*.

S. Steinerberger, *The Unreasonable Elegance of Mathematics*,
Springer Undergraduate Mathematics Series, https://doi.org/10.1007/978-3-032-03815-9_19

It is a curious and little known fact that one of the intellectual predecessors was Erasmus Darwin, Charles Darwin's very own *grandfather*, who wrote

> would it be too bold to imagine, that in the great length of time, since the earth began to exist, perhaps millions of ages before the commencement of the history of mankind, would it be too bold to imagine, that all warm-blooded animals have arisen from one living filament, which THE GREAT FIRST CAUSE endued with animality, with the power of acquiring new parts, attended with new propensities, directed by irritations, sensations, volitions, and associations; and thus possessing the faculty of continuing to improve by its own inherent activity, and of delivering down those improvements by generation to its posterity, world without end! (Erasmus Darwin, *Zoonomia or The Laws of Organic Life* [69])

Erasmus Darwin's question, *would it be too bold to imagine, that all warm-blooded animals have arisen from one living filament*, is echoed by his grandson Charles, who argues that *I should infer from analogy that probably all the organic beings which have ever lived on this earth have descended from some one primordial form, into which life was first breathed* [68]. Carefully mentioning THE FIRST GREAT CAUSE did not help Erasmus, *Zoonomia* promptly landed on the Vatican's Index of Forbidden Books (an honor that his grandson Charles would never be granted).

84   **INDEX LIBRORUM**
Darwin Erasmo Medico di Derby, Membro etc. Zoonomia,
ovvero Leggi della Vita Organica. Traduzione dall'

Fig. 19.1: Erasmus Darwin in the Index of Forbidden Books (1819): *Erasmus Darwin, Doctor in Derby, Member etc. Zoonomia or The Laws of Organic Life*

In the spirit of intellectual predecessors, Freud is clear that when it comes to psychological humiliation, he was not the first point it out: *Let us hasten to add that Psychoanalysis was not the first to take this step. Famous philosophers preceed us, especially the great thinker Schopenhauer* [92]. Being human is humbling!

19.2 Hilbert's Program

Suppose we have now accepted the first three humiliations, we are no longer at the center of the universe, we are not all that different from some of our animal friends, and we are not even entirely certain what is happening inside ourselves. So far, so good. However, even in the darkest moments, we still have mathematics – and mathematics is certain! It is so certain that people even copy the way it looks!

> One should not be deceived by philosophical works that pretend to be mathematical, but are merely dubious and murky metaphysics. Just because a philosopher can recite the words Lemma, Theorem and Corollary doesn't mean that his work has the certainty of mathematics. That certainty does not derive from big words [...] but rather from the utter simplicity of the objects considered by mathematics. (Maupertuis [189])

So let us focus on developing mathematics further. We already saw that an overly simplistic use of mathematics can lead to contradictions – and when it comes to mathematics, we should really try to be sure. As David Hilbert proclaimed in a 1930 lecture *A science like mathematics must not rest on faith, no matter how strong this faith, it must rest on the duty of complete resolution* [133]. That is a worthwhile endeavor! Hilbert urges his colleagues to strictly formalize everything: *Everything that constitutes mathematics will be strictly formalized so that mathematics will become a collection of formulas. These [formulas] will be different from the usual ones only insofar as, except for the usual symbols, it also contains the logical symbols 'implies' ($\Longrightarrow$) and 'not' ($\neg$). Some formulas, which will serve as the foundations, are called axioms. A proof [...] consists of steps where every assumption is either an Axiom or coincides with the conclusion of a previous argument* [133]. At this point, this is probably already familiar to the reader. Hilbert then goes further and says

> Our most important duty is now to prove the following [...]: 1. If a statement does not cause contradictions, then it can also be proved. (Hilbert [133])

There are two things packaged into one: if a statement is true, then its negation is false (and will contradict other statements). If a statement does not contradict anything, it must be true *and we can prove it.* Hilbert is not alone in thinking this, Wittgenstein, just a few years earlier, wrote *If a question can be put at all, then it can also be answered* (Tractatus 6.5). In a radio address, also given in 1930, Hilbert goes so far as to say that *We must not believe those who, with philosophical airs and superior tone, prophecy the end of all culture and who revel in the Ignorabimus* [We will not know]. *For us [mathematicians] there is no Ignorabimus, and in my opinion it does not exist in the natural sciences. Instead of the foolish Ignorabimus, our solution may be: We have to know, we will know.* David Hilbert lived by this idea and is now a strong contender for having one of the most powerful tombstone inscriptions of all time: DAVID HILBERT. WE HAVE TO KNOW. WE WILL KNOW.

19.3 The Fourth Humiliation

At the very same conference where Hilbert made the above remarks (the *Second Conference on the Epistemology of the Exact Sciences* held in Königsberg from 5-7 September 1930), the young Austrian logician Kurt Gödel (1906–1978) briefly presented some ideas that would deal a fatal blow to Hilbert's philosophical program. Gödel's First Incompleteness Theorem not only dooms the entirety of Hilbert's program but also puts a fundamental limitation to what we (or *any* intelligent being) can hope to achieve. It says, simply put, that every time we argue within an axiomatic system, there will be statements within the system whose truth we cannot decide within the system. There are true statements that cannot be proven – not because they are very difficult and complicated or because they exceed our limited abilities, but because it is in their very nature.

Zermelo was a very irascible person. He had suffered a nervous breakdown and felt ill-treated, but had actually recovered at that time. He had no wish to meet Gödel. A small group suggested lunch at the top of a nearby hill, which involved a mild climb, with the idea that Zermelo should talk to Gödel. Zermelo did not want to do so and made excuses: he did not like Gödel's looks (he actually had not met him); the climb was too much for him; there would not be enough food if Gödel came along too. When Gödel joined the group, however, the two immediately started discussing logic, and Zermelo never noticed that he had made the climb. (Olga Taussky-Todd, *Remembrances of Kurt Gödel*, [276])

Fig. 19.2: Kurt Gödel (left) and Olga Taussky-Todd remembering the meeting between Gödel and the logician Ernst Zermelo.

One of the people who attended that meeting was John von Neumann, who immediately proposed to make it more concrete. *Von Neumann was very enthusiastic about the result and had a private discussion with Gödel. In this discussion, von Neumann asked whether number-theoretical undecidable propositions could also be constructed in view of the fact that the combinatorial objects can be mapped onto the integers and expressed the belief that it could be done. In reply, Gödel said, "Of course undecidable propositions about integers could be so constructed, but they would contain concepts quite different from those occurring in number theory like addition and multiplication". Shortly afterward Gödel, to his own astonishment, succeeded in turning the undecidable proposition into a polynomial form preceded by quantifiers* [292].

> **The Fourth Humiliation.** Assuming our standard set of axioms, there are true statements about the integers that we are unable to prove. The set of *true statements* is strictly larger than the set of *provable statements*. Some truths will remain elusive forever.

Maybe there is something wrong with our axioms? That is not where the problem lies! No matter what axioms one chooses, if the arising formal system is sufficiently nontrivial for us to be able to do a sufficient amount of elementary arithmetic, then there will be such true but unprovable statements. It is not a problem with the axioms, it is a fundamental limitation of what one can hope to achieve. Any formal system brings with it its own limitations. More generally, formal systems create their own undecidable statements: these are statements that may sound perfectly legitimate but whose truth or falsehood is actually undecidable within the framework that we are given. A particularly shocking example of such an undecidable statement is the *continuum hypothesis*. We have seen above that $\mathbb{N}$ and $\mathbb{Z}$ have the same cardinality: there exists a bijective map $f : \mathbb{N} \to \mathbb{Z}$ and thus they have the same cardinality: the two sets have, in the Cantorian sense, the same 'size'. We have also seen, as a consequence of the Cantor Diagonal Argument that any function $f : \mathbb{N} \to \mathbb{R}$ has to necessarily 'miss' some real numbers, there exists $x \in \mathbb{R}$ such that for all $n \in \mathbb{N}$ one has $f(n) \neq x$. We interpret this as saying that $\mathbb{R}$ is strictly larger than $\mathbb{N}$.

Question. Is there a set X such that X is strictly larger than $\mathbb{N}$ and strictly smaller than $\mathbb{R}$?

This question looks potentially difficult, but it does not seem to be philosophically challenging: either such a set exists or not. In general, every question starting with 'Is there...' sounds like it should have a simple answer: yes or no. It may therefore come as a surprise that this question is actually *undecidable* within the current standard framework of mathematics. You may assume that such a set X exists and proceed from there, and you will never run into any contradictions with the existing mathematical framework. You may likewise assume that such a set does not exist, and that assumption will also never lead to any contradictions. If you find this somewhat unsettling, you are not alone. A few weeks after the Königsberg conference, von Neumann sent a letter to Gödel that starts with another shock!

> Dear Mr. Gödel!   (Berlin, 20. November 1930)
> I have recently concerned myself again with logic, using the methods you have employed
> so successfully in order to exhibit undecidable properties. In doing so I achieved a result
> that seems to me to be remarkable. Namely, I was able to show that the consistency of
> mathematics is unprovable. [...] And again: aren't we going to see you soon in Berlin?
> E. Schmidt, to whom I communicated your result as you had presented it in Königsberg,
> was delighted by it. He considers it, as I do, to be the greatest logical discovery in a long
> time. (von Neumann to Gödel [110])

We cannot even be sure that there are no contradictions! What must have come as another surprise to von Neumann is Gödel's reply who informed von Neumann that he himself had already discovered the same result and sent it for publication. Von Neumann, a true scholar, immediately replies *Many thanks for your letter and your reprint. As you have established the theorem on the unprovability of consistency as a natural continuation and deepening of your earlier results, I clearly won't publish on this subject* [110]. In summary, there are true statements we cannot prove, and it is not even entirely clear that what we have done so far is actually self-consistent in the sense of not eventually leading to contradictions. It is a difficult way to live.

> God exists, since mathematics is consistent, and the Devil exists, since we cannot prove it.
> (attributed to André Weil [245])

The philosopher Paul Benacerraf [24] even says *Gödel, however, is the missing link, for he supposedly proved, very roughly speaking, that if mathematics is consistent we cannot prove it. It might therefore be said that he clinched the case for Satan's existence. [...] These are heady matters, dark doings*[1]*, and I do not propose to discourse the present state of mathematical theology.*

19.4 So where does this leave us?

At this point, we know that there are statements that are true but not provable. What are they like? In some sense, we will never know: a statement which is true but

---

[1] Regarding Gödel's *dark doings* it is worth noting that *judged by the amount of space in Gödel's note books dealing with general philosophy and theology, including demonology, these subjects occupied a great deal of his attention ever since his student days* [159].

cannot be proven is indistinguishable from a statement that is true but which we cannot currently prove (and there are many of those!). Some people believe that such unprovable statements must necessarily be very complicated and possibly not very interesting: fundamental results, like the Pythagorean Theorem, are so deeply embedded in the nature of mathematics that we will always be able to understand why they are true. Others disagree.

> It has long been known that there are arithmetic statements that are true but not provable, but it is usually thought that they must necessarily be complicated. In this paper, I shall argue that these wild beasts may be just around the corner. (John Conway [60])

Conway proposes the following iteration game, very much inspired by the $3x + 1$ Collatz conjecture: each number n can be written as $n = 2k$ or $n = 4k - 1$ or $n = 4k + 1$, the iteration rule proposed by Conway is

$$2k \to 3k \qquad 4k + 1 \to 3k + 1 \qquad \text{and} \qquad 4k - 1 \to 3k - 1.$$

He calls it the *amusical permutation* $\mu(n)$ and notes that μ is a permutation (and thus so is the inverse μ^{-1}).

> *The simplest assertion about μ that I believe to be true but unsettleable is that 8 belongs to an infinite cycle. [...] this convinced me that the Collatz $3n + 1$ Conjecture is itself very likely to be unsettleable.* (John Conway [60])

There are also very concrete examples of our limitations. A Diophantine equation is a polynomial equation with a finite number of unknowns and integer coefficients, examples being $x^2 + y^2 = z^2$ or $x^4 + y^2 = -8$. Some Diophantine equations have a solution involving only integers, for example, $3^2 + 4^2 = 5^2$, while others do not. Hilbert [131] asked, in 1900, whether there is a general procedure that can help us decide, for any such equation, whether it has an integer solution.

> **X. — De la possibilité de résoudre une équation de Diophante.**
>
> On donne une équation de Diophante à un nombre quelconque d'inconnues et à coefficients entiers rationnels : *On demande de trouver une méthode par laquelle, au moyen d'un nombre fini d'opérations, on pourra distinguer si l'équation est résoluble en nombres entiers rationnels.*

10. Determination of the Solvability of a Diophantine Equation. Given a Diophantine equation with any number of unknown quantities and with rational integral numerical coefficients: to devise a process according to which it can be determined in a finite number of operations whether the equation is solvable in rational integers. (David Hilbert [132])

It is a result of Matiyasevich, Robinson, Davis and Putnam that this is *not* possible. A magic oracle cannot exist. Well, life is certainly more interesting that way!

Acknowledgment

if I understood right, you are discovering that librarians know something useful
(Dr. Johanna Wilhelmina Smit, former Director of the General
Archive of the Universidade de São Paulo, and mother-in-law)

This project would have been impossible if it were not for the great libraries and fantastic librarians of the world. *Chapter 2.* Figures 2.2 and 2.4 are courtesy of The Linda Hall Library of Science, Engineering & Technology under CC by 4.0. Fig. 2.3. was generously provided by The British Library [C.175.i.41.] (with very special thanks to Katerina Borisova and Elias Mazzucco) as well as the Austrian National Library (with special thanks to Dr. Gertrud Oswald). Fig. 2.5 was provided by Numdam (special thanks to Céline Smith). Fig. 2.6 was generously made available by The British Library, which has a representation of the original manuscript as Microfilm Reel #338, ESTC Number T110172 (with very special thanks to Sandra Powlette). Fig. 2.9 has been made public by the Library of Congress, special thanks to the Lessing J. Rosenwald Collection. *Chapter 3.* Comic courtesy of xkcd.com (special thanks to Casey Blair). *Chapter 4.* Fig. 4.5 was made possible by the generosity of Prof. James St. John. *Chapter 7.* Fig. 7.2 was made possible by e-rara [https://doi.org/10.3931/e-rara-8490], special thanks to Dr. Meda Diana Hotea and the ETH Library. I also acknowledge help from Frank Vanlangenhove and Femke Van der Fraenen from the Universiteitsbibliotheek Ghent, which houses another copy [BIB.CL.000781/1]. *Chapter 8.* Fig. 8.2. was generously made available by the Pontifical Institute of Mediaeval Studies Library at the University of Toronto (special thanks to Dr. Greti Dinkova-Bruun). Fig. 8.4, like scans of all of Euler's work, have been made available by the Euler Archive at https://scholarlycommons.pacific.edu/euler/ (special thanks to Prof. Erik R. Tou). Fig. 8.5 is courtesy of the John Carter Brown Library, special thanks to Kim Nusco (additional thanks to the Wellcome Collection and Odalys Caballero). Fig. 8.6 is courtesy the ARTFL Encyclopédie Project, University of Chicago (special thanks to Clovis Gladstone). Fig. 8.7 was made available by the Boston Public Library (special thanks to John J. Devine). *Chapter 11.* Fig. 11.2 was made possible by the Bavarian State Library (special thanks to Sylvia Herfurth). Fig. 11.3 was generously made available by the George Fabyan Collection, Rare Book & Special Collections Division, Library of Congress. *Chapter 13.* The wonderful Fig. 13.1 was produced by Dall-E. Fig. 13.4 was generously provided by the Bibliothèque nationale de France (many thanks to Eric Vigneron). Fig. 13.5 was graciously provided by Springer Nature. *Chapter 15.* Fig. 15.1 is courtesy of The Linda Hall Library of Science, Engineering & Technology under CC by 4.0. The picture of Gruenberger's article was generously provided by RAND (special thanks to Cindy Lyons). *Chapter 16.* Fig. 16.2 is provided by the Smithsonian Libraries and Archives [https://doi.org/10.5479/sil.262971.39088000323931]. Fig. 16.4 was generously provided by the Bibliothèque nationale de France (many thanks again to Eric Vigneron). Figures 16.7, 16.8 and 16.12 were provided by Numdam (thanks again to Céline Smith). Fig. 16.10 was generously provided by Springer Nature. *Chapter 17.* Fig. 17.1 was provided by the Smithsonian Libraries and Archives as well as the French Academy of Science. *Chapter 18.* Fig. 18.1 was made possibly by the generosity of Ryan North (beware of cephalopod neighbors!). *Chapter 19.* Fig. 19.1 was provided by the Boston Public Library (special thanks to John J. Devine). *Additional thanks to* the National Gallery of Victoria, Melbourne (especially Eleanor Jordan-Gahan and Philip White).

This project would have been many orders of magnitude more complicated were it not for the University of Washington Library. Its Mathematics Research Library and the Rare Books Collection have been a blessing. They helped with the remaining figures, access to rare books and wonderful easy-to-use scanning equipment. I am especially grateful to Maryam Fakouri who was tremendously helpful and a constant source of wisdom.

References

1. G. B. Airy, *Monthly Notices of the Royal Astronomical Society*, Volume 7, Issue 9, November 1846, Pages 121–144.

2. Jean le Rond d'Alembert, *Opuscules Mathématiques*, vol. 2, Paris 1761, Quatorzieme Memoire, p. 239.

3. S. Alexander, Price movements in speculative markets: Trends or random walks, *Industrial Management Review* (1961), **2**(2), 7–26.

4. M. Allais, Le comportement de l'homme rationnel devant le risque: critique des postulats et axiomes de l'ecole Americaine, *Econometrica* **21**(4) (1953), 503–546.

5. S. Andreski, *Social sciences as sorcery*, London, UK: Andre Deutsch, 1972.

6. R. Apéry, Irrationalité de $\zeta(2)$ et $\zeta(3)$, *Astérisque* **61** (1979), 11–13.

7. Archimedes, The Works of Archimedes, edited by Sir Thomas Little Heath, University of Michigan University Press, 1897.

8. Archimedes, *Opera omnia*, edited by Johan Ludvig Heiberg, Cambridge University Press, 1881

9. Aristotle, *Physics*, translated by R. P. Hardie & R. K. Gaye, *The Works of Aristotle*, Clarendon Press, Oxford, 1930.

10. M. Armgardt, Leibniz as legal scholar, *Fundamina: A Journal of Legal History* **20**(1) (2014), 27–38.

11. Antoine Arnauld and Pierre Nicole, *Logic, or the Art of Thinking being The Port-Royal Logic*, translated from the French with an introduction by Thomas Spencer Baynes, Sutherland and Knox, Edinburgh, 1850.

12. Tribute to Vladimir Arnold, edited by B. Khesin and S. Tabachnikov, *Notices of the American Mathematical Society*, 2012.

13. V. I. Arnol'd, On teaching mathematics, *Russ. Math. Surv.* **53** (1998), 229–236.

14. R. Axelrod, *The Evolution of Cooperation*, Basic Books, 1984.

15. L. Bachelier, Theorie de la speculation, *Annales scientifiques de l'E.N.S.* 3e serie, tome 17 (1900), 21–86.

16. Bacon, Francis, William Rawley, and George Fabyan Collection, *Sylva sylvarum, or, A natural history in ten centuries: together with the History natural and experimental of life and death, or of the prolongation of life: whereunto is added Articles of enquiry touching metals and minerals and the New Atlantis, with an alphabetical table of the principal things contained in the ten centuries*, London: Printed for Bennet Griffin, 1683. Library of Congress, 95202443.

17. Roger Bacon, *The Opus Majus of Roger Bacon*, translated by Robert Belle Burke, University of Pennsylvania Press, 1928.

18. Roger Bacon, *The Famous Historie of Fryer Bacon*, edited by Edmund Goldsmid, privately printed, Edinburgh, 1886.

19. L. Badger, Lazzarini's lucky approximation of π, *Mathematics Magazine* **67** (2) (1994), 83–91.

20. D. Bailey, J. M. Borwein, R. Crandall, and C. Pomerance, On the binary expansions of algebraic numbers, *J. Theor. Nombres Bordeaux* **16**(3) (2004), 487–518.

21. Illustration from James Baldwin, *Thirty more famous stories retold*, Vol. 481. American Book Company, 1905.

22. B. Barbiellini-Amidei, The Casimir effect in conformal field theories, *Physics Letters B* **190** (1987), 137–139.

23. J. Bell and V. Blasjo, Pietro Mengoli's 1650 Proof that the Harmonic Series Diverges, *Mathematics Magazine* **91** (2018), 341–347.

24. P. Benacerraf, God, the Devil and Gödel, *The Monist* **51** (1967), 9–32.

25. Isaiah Berlin in Conversation with Bryan Magee, *An Introduction to Philosophy*, Men of Ideas, BBC Television Show, 1978

26. B. Berndt and R. Rankin, *Ramanujan: Letters and Commentary*, American Mathematical Society and London Mathematical Society, 1995.

© The Editor(s) (if applicable) and The Author(s), under exclusive license to Springer Nature Switzerland AG 2025

S. Steinerberger, *The Unreasonable Elegance of Mathematics*, Springer Undergraduate Mathematics Series, https://doi.org/10.1007/978-3-032-03815-9

27. Daniel Bernoulli, originally published in 1738 ("Specimen Theorize Naval de Mensura Sortis", "Commentarii Academiae Scientiarum Imperialis Petropolitanae"); translated by Dr. Louise Sommer (January 1954). "Exposition of a New Theory on the Measurement of Risk", *Econometrica* **22**(1), 22–36.

28. *The King James Bible*, 1611

29. I.-J. Bienaymé, Considérations àl'appui de la découverte de Laplace, *Comptes Rendus de l'Acadmie des Sciences* **37** (1853), 309–324.

30. F. Black and M. Scholes, The Pricing of Options and Corporate Liabilities, *Journal of Political Economy*, Vol. 81, No. 3 (1973), 637–654

31. F. H. Bool, J. R. Kist, J. L. Locher, and F. Wierda, *M. C. Escher: His Life and Complete Graphic Work*, Harry N. Abrams, New York, l982; Abradale Press, 1992.

32. E. Borel, Les probabilités dénombrables et leurs applications arithmétiques, *Rendiconti del Circolo Matematico di Palermo* **27** (1909), 247–271.

33. James Boswell, *Boswell's Life of Johnson*, Abridged and edited, with an introduction by Charles Grosvenor Osgood, Project Gutenberg.

34. Louis-Antoine Fauvelet de Bourrienne, *Memoires de M. de Bourrienne sur Napoleon, le Directoire, le Consulat, l'Empire et la Restauration*, Paris, Ladvocat, 1829–1831.

35. Carl Boyer, *The History of the Calculus and Its Conceptual Development*, Dover Books on Mathematics, 1959.

36. Ernst Breitenberger, Gauss's Geodesy and the Axiom of Parallels, *Archive for History of Exact Sciences*, Vol. 31, No. 3 (1984), 273–289.

37. E. Bretscher and J. D. Cockcroft, Enrico Fermi 1901–1954, *Biographical Memoirs of Fellows of the Royal Society* **1** (1955), 69–78.

38. H. Brocard, Question 1532, *Nouv. Ann. Math.* **4** (1885), 391.

39. P. Brosche, Die Wiederauffindung der Ceres im Jahre 1801, *Acta Historica Astronomiae*, vol. 14 (2002), 80–88.

40. *The Selected Correspondence of L. E. J. Brouwer*, edited by Dirk van Dalen, Springer, 2011.

41. N. J. L. Brown, A. D. Sokal, H. L. Friedman, The complex dynamics of wishful thinking: The critical positivity ratio, *American Psychologist* **68**(9) (2013), 801–813.

42. R. Burton, The Anatomy of Melancholy, *New York Review Books*, 2001.

43. G. Cantor, Ueber eine elementare Frage der Mannigfaltigkeitslehre, *Jahresbericht der Deutschen Mathematiker-Vereinigung* **1** (1891), 75–78.

44. S. Caracciolo, M. D'Achille, V. Erba and A. Sportiello, The Dyck bound in the concave 1-dimensional random assignment model, *Journal of Physics A: Mathematical and Theoretical* **53**(6) (2020), 064001.

45. Girolamo Cardano, *The Book of My Life (De Vita Propria Liber)*, translated by Jean Stoner, E. P. Dutton & Co, Inc.

46. Girolamo Cardano, *The Book on Games of Chance (Liber de Ludo Aleae)*, translated by Sydney Henry Gould, in: Oystein Ore, Cardano: *The Gambling Scholar*, Princeton University Press, 2017.

47. Girolamo Cardano, *De Subtilitate libri XXI*, Lugduni Lyon: Apud Stephanum Michaelem, 1580, courtesy of the John Carter Brown Library.

48. Girolamo Cardano, *De Subtiitate*, Edited by John M. Forrester, with an Introduction by John Henry and John M. Forrester, Arizona Center for Medieval and Renaissance Studies, Tempe, Arizona, 2013.

49. Girolamo Cardano, *Hieronymi Cardani mediolanensis opera omnia, t.1 quo continentur philologica, logica, moralia*, Lugduni: Ioannis Antonii Huguetan & Marci Antonii Ravaud, 1663, scan made public by Digital Library dell'Universita degli Studi di Torino.

50. J. S. Chahal, Pell's Equation and the Unity of Mathematics, *History of the Mathematical Sciences* **2** (2004), 163–170.

51. D. Champernowne, The construction of decimals normal in the scale of ten, *Journal of the London Mathematical Society* **8** (1933), 254–260.

52. S. Chapman, XXXV. On ozone and atomic oxygen in the upper atmosphere, *The London, Edinburgh, and Dublin Philosophical Magazine and Journal of Science* 10.64 (1930), 369–383.

53. G. Chartrand, T. Haynes, S. Hedetniemi and P. Zhang, Editorial Remembering Frank Harary, *Discrete Mathematics Letters* **6** (2021), 1–7.

54. P. Chebychev, Des valeurs moyennes, *Journal de mathématiques pures et appliquées*, 2e série, tome 12 (1867), 177–184.

55. T. Chiang, *Stories of your life and others*, Knopf, 2010.

56. Cicero, *The Letters of Cicero; the whole extant correspondence in chronological order, in four volumes*, Evelyn S. Shuckburgh. London. George Bell and Sons 1908–1909.

57. Cicero, *Tusculan Disputations, also, Treatise on the Nature of the Gods and on the Commonwealth*, translated chiefly by C. D. Yonge, Harper & Brothers, New York, 1877.

58. J. P. Cleave, Cauchy, Convergence and Continuity, *The British Journal for the Philosophy of Science*, Vol. 22, No. 1 (Feb., 1971), 27–37.

59. T. Cohen and W. Knight, Convergence and Divergence of $\sum_{n=1}^{\infty} 1/n^p$, *Mathematics Magazine* **52** (1979), 178.

60. J. H. Conway, On Unsettleable Arithmetical Problems, *The American Mathematical Monthly* **120** (2013), 192–198.

61. M. Cook, Universality in elementary cellular automata, *Complex systems* **15** (2005), 1–40.

62. A. Cournot, *Recherches sur les principes mathématiques de la théorie des richesses*, L. Hachette (Paris), 1838

63. H. S. M. Coxeter, *Regular polytopes*, Dover Publication, Inc, 1973.

64. H. S. M. Coxeter, M. Emmer, R. Penrose, and M. L. Teuber, eds. *M. C. Escher: Art and Science*, North-Holland, Amsterdam, 1986.

65. Nicolai de Cusa, *Opera omnia iussu et auctoritate Academiae litterarum heidelbergensis ad codicum fidem edita*, Volumes 4-5, in aedibus Felicis Meiner, 1959

66. N. Cutland, C. Kessler, E. Kopp and D. Ross, On Cauchy's Notion of Infinitesimal, *The British Journal for the Philosophy of Science*, Vol. 39, No. 3 (Sep., 1988), 375–378.

67. Dante, *The Divine Comedy*, translated by the Rev. Henry Francis Cary, Cassell & Company, London.

68. C. Darwin, *On the origin of species by means of natural selection, or the preservation of favoured races in the struggle for life*, London: John Murray, 1859.

69. E. Darwin, *Zoonomia or The laws of organic life*, Boston: Thomas and Andrews, 1809

70. J. Dauben, *Georg Cantor: His Mathematics and Philosophy of the Infinite*, Princeton University Press, 1990

71. L. Downing, The Prisoner's Dilemma Game as a Problem-Solving Phenomenon: An Outcome Maximizing Interpretation, *Simulation and Games* **6** (1975), 366–91.

72. S. Drake, *Galileo Studies: Personality, Tradition and Revolution*, The University of Michigan Press, Ann Arbor, 1970.

73. L. Dubins and L. Savage, *How to Gamble If You Must: Inequalities for Stochastic Processes*, Dover, 1965.

74. J. Dupuis, *Le Nombre Geometrique de Platon*, Paris: Hachette, 1885.

75. W. Durant, *The Story of Civilization Volume 9: The Age of Voltaire*, Simon & Schuster, 1967.

76. C. W. Dyck, *Wolff and the First Fifty Years of German Metaphysics*, preprint available at philpapers.org.

77. R. Eckhardt, Stan Ulam, John Von Neumann, and the Monte Carlo Method, *Los Alamos Science* **15** (1987), 131–137.

78. A. Eddington, *The nature of the physical world*, Cambridge University Press, 1929.

79. Andreas von Ettingshausen, *Die combinatorische Analysis als Vorbereitungslehre zum Studium der theoretischen höhern Mathematik*, J.B. Wallishausser, 1826.

80. L. Euler, Variae observationes circa series infinitas, *Commentarii academiae scientiarum Petropolitanae* **9** (1744), 160–188.

81. L. Euler, Remarques sur un beau rapport entre les séries des puissances tant directes que réciproques, *Mémoires de l'Académie des Sciences de Berlin*, Volume 17, 1768, 83–106. Original text available online from the Euler Archive, at https://scholarlycommons.pacific.edu/euler/.

82. P. Feyerabend, *Against method: Outline of an anarchistic theory of knowledge*, Verso Books, 2020.

83. R. Feynman, *Surely you're joking, Mr. Feynman!: adventures of a curious character*, WW Norton & Company, 2010.

84. J. G. Fichte, *The Vocation of Man*, J. Chapman, 1848.

85. H. Fischer, *A history of the central limit theorem: from classical to modern probability theory* (Vol. 4). New York: Springer, 2011.

86. E. Forbes, Gauss and the Discovery of Ceres, *Journal for the History of Astronomy* vol. 2, 1971, p. 195.

87. B. Franklin, *Benjamin Franklin: His Autobiography*, Derby & Jackson, 1859.

88. B. Fredrickson and M. Losada, Positive affect and the complex dynamics of human flourishing, *American psychologist* **60** (2005), 678–686.

89. B. Fredrickson, Updated thinking on positivity ratios, *American Psychologist* **68** (2013), 814–22.

90. Gottlob Frege, *Philosophical and Mathematical Correspondence*, Abridged from the German edition by B. McGuinness. Translated by H. Kaal. Oxford: Basil Blackwell; Chicago: University of Chicago Press, 1980.

91. G. Frege, *The Basic Laws of Arithmetic*, ed. and trans. M. Furth (Berkeley: Univ. of California Press, 1964), p. 127.

92. S. Freud, Eine Schwierigkeit der Psychoanalyse. In: *Imago. Zeitschrift für Anwendung der Psychoanalyse auf die Geisteswissenschaften*, Bd. V (1917). S. 1–7.

93. H. Friedman, *Philosophical Problems in Logic*, Seminars presented at Princeton, 2002.

94. H. Frost, *Lilybaeum (Marsala) – The Punic Ship: Final Excavation Report*, Notizie degli scavi di antichita. 8. XXX. Rome: Accademia Nazionale dei Lincei, 1981.

95. D. Gale and L. S. Shapley, College admissions and the stability of marriage, *The American Mathematical Monthly* **69**(1) (1962), 9–15.

96. Galileo Galilei, *Discourses and Mathematical Demonstrations Relating to Two New Sciences*, translated by Henry Crew and Alfonso de Salvio, Macmillan, 1914.

97. Galileo Galilei, *Le Opere di Galileo Galilei - Edizione nazionale* (Volume 13), p. 387, Opere, VIII, Tip. di G. Barbera, Firenze, 1890

98. F. Galton, *The art of travel, or, Shifts and contrivances available in wild countries*, London: John Murray, 1855.

99. F. Galton, Statistical Inquiries into the Efficacy of Prayer, *Fortnightly Review*, vol. 12, pp. 125–35, 1872.

100. F. Galton, Regression towards mediocrity in hereditary stature, *The Journal of the Anthropological Institute of Great Britain and Ireland* **15** (1886), 246–263.

101. Francis Galton, *Natural Inheritance*, New York; Macmillan and co., 1894.

102. F. Galton, Cutting a Round Cake on Scientific Principles (Letters to the Editor), *Nature*, 20 December 1906.

103. W. Gangbo and R. McCann, The geometry of optimal transportation, *Acta Math.* **177** (1996), 113–161.

104. James Garfield, Pons Asinorum, *New England Journal of Education*, Vol. 3 (1876), p. 161.

105. P. Gassendi, *Syntagma*, in *Opera omnia: in sex tomos divisa*, Vol. I, Sumptibus L. Anisson, & I.B. Devenet, 1658.

106. C. F. Gauss, *Theory of the Motion of the Heavenly Bodies moving about the sun in conic sections*, a translation of Gauss's 'Theoria Motus' by Charles Henry Davis, Little, Brown and Company, Boston, 1857.

107. C. F. Gauss and F. W. Bessel, *Briefwechsel zwischen Gauss und Bessel*, W. Engelmann, 1880.

108. C. F. Gauss and H. C. Schumacher, *Briefwechsel*, vol 4, Gustav Esch, 1862.

109. K. Gödel, Über formal unentscheidbare Sätze der Principia Mathematica und verwandter Systeme I, *Monatshefte der Mathematik* **38** (1931), 173–198.

110. K. Gödel: *Collected Works: Volume V*, edited by Solomon Feferman, John W. Dawson Jr, Warren Goldfarb, Charles Parsons, Wilfried Sieg, Oxford University Press, 2013.

111. Kurt Gödel, What is Cantor's continuum problem?, in P. Benacerraf and H. Putnam (eds.), *Philosophy of Mathematics*, 2nd edition (Cambridge: Cambridge University Press, 1983), pp. 483–84.

112. J. W. v. Goethe, *Faust: A tragedy by Joh. Wolfg. von Goethe*, translated, in the original metres, by Bayard Taylor. I. United Kingdom: Strakan & Company.

113. S. W. Golomb, Checker Boards and Polyominoes, *Amer. Math. Monthly* **61** (1954), 675–682.

114. I. J. Good, Speculations Concerning the First Ultraintelligent Machine, *Advances in computers*, Vol. 6 (1966), 31–88.

115. R. L. Graham and H. O. Pollak, Note on a nonlinear recurrence related to $\sqrt{2}$, *Math. Mag.* **43** (1970), 143–145.

116. I. Grattan-Guinness, Towards a biography of Georg Cantor, *Annals of Science* **27**(4) (1971), 345–391.

117. R. Grosseteste, *De Luce*, translated with an introduction by Clare C. Riedl, M. A., Marquette University Press, 1942.

118. B. Grünbaum, Is Napoleon's Theorem Really Napoleon's Theorem?, *American Mathematical Monthly* **119** (2012), 495–501.

119. F. Gruenberger, A Measure for Crackpots, *RAND Memorandum*, 1962.

120. F. Gruenberger, A Measure for Crackpots: How does one distinguish between valid scientific work and counterfeit "science"?, *Science* 145.3639 (1964), 1413–1415.

121. B. Gustafsson, *Scientific Computing: A Historical Perspective*, Texts in Computational Science and Engineering, Vol. 17, 2018, Springer.

122. Ian Hacking, *The emergence of probability: A philosophical study of early ideas about probability, induction and statistical inference*, Cambridge University Press, 2006.

123. P. Halmos, Von Neumann on measure and ergodic theory, *Bulletin of the American Mathematical Society* **64**(3) (1958), 86–94.

124. G. H. Hardy, S. Ramanujan, FRS. *Nature* **105** (1920), 494–495.

125. G. H. Hardy, *An Annotated Mathematician's Apology*, annotated by Alan J. Cain, 2024.

126. G. H. Hardy, *Divergent series*, (Vol. 334), American Mathematical Society, 2024.

127. John C. Harsanyi, Cardinal Utility in Welfare Economics and in the Theory of Risk-taking, *Journal of Political Economy* **61**(Oct., 1953), 434–435.

128. J. van Heijenoort, ed., *From Frege to Gödel*, Cambridge, Mass.: Harvard Univ. Press, 1967.

129. J. Van Heijenoort, ed., *From Frege to Gödel: A source book in mathematical logic, 1879–1931*, Harvard University Press, 1999.

130. R. Henriques, The Kundmanngasse and Its Significance. In *Self-understanding in the Tractatus and Wittgenstein's Architecture: From Adolf Loos to the Resolute Reading* (pp. 1–32). Cham: Springer Nature Switzerland, 2024.

131. D. Hilbert, *Compte Rendu du Deuxieme Congres International des Mathematiciens, tenu a Paris du 6 a 12 Aout 1900*, Gauthier-Villars, Imprimeur-Libraire du bureau des longitudes, de l'Ecole Polytechnique, 1902.

132. D. Hilbert, Mathematical problems, *Bull. Amer. Math. Soc.* **8**(10) (1902), 437–479.

133. D. Hilbert, Die Grundlegung der elementaren Zahlenlehre, *Mathematische Annalen* **104** (1931), 485–494.

134. T. Hobbes, *Leviathan*, in *The English Works of Thomas Hobbes*, ed. William Molesworth (London, 1845), rpt. (Darmstadt: Scientia Verlag, 1966), Vol. III, 17 (idea of infinity).

135. T. Hobbes, *De principiis & ratiocinatione geometrarum ubi oftenditur incertitudinem falsitatemq; non minorem inesse scriptis eorum, quam scriptis physicorum & ethicorum. / Contra fastun professorum geometriae*, London: Apud Andream Crooke in Coemiterio D. Pauli sub signo Draconis viridis, 1666.

136. *The English Works of Thomas Hobbes of Malmesbury; now first collected and edited by Sir William Molesworth, Bart.* Volume 7. London: Longman, Brown, Green and Longmans, 1845.

137. T. Hobbes, *De Cive. Philosophicall Rudiments Concerning Government and Society etc.*, London, Printed by J.C. for R. Royston, at the Angel in Ivie-Lane, 1651.

138. Paul Hoffman, *The Man Who Loved Only Numbers*, Fourth Estate, London, 1999.

139. N. N. Holland, Freud on Shakespeare, *PMLA*, Vol. 75, No. 3 (Jun., 1960), 163–173.

140. Homer, *The Iliad of Homer*, Translated by Alexander Pope, With Notes and Introduction by the Rev. Theodore Alois Buckley, M.A., F.S.A. and Flaxman's Designs, 1899.

141. Homer, *The Odyssey*, translated by Alexander Pope, Bernard Lintot, London, 1725.

142. Homer, *The Odyssey*, rendered into English prose for the use of those who cannot read the original by Samuel Butler, Longmans, Green and Co, London, 1900.

143. C. Huygens, *De Ratiociniis in Ludo Aleae or The Value of all Chances in Games of Fortune*, Printed by S. Keimer for T. Woodward, near the Inner-Temple-Gate in Fleetstreet, 1714.

144. *Index librorum prohibitorum*, Typis Reu. Cam. Apost, 1704.

145. Pope Innocent IV, Papal Bull *Ad extirpanda*, May 15, 1252.

146. D. Jesseph, Geometry, religion and politics: context and consequences of the Hobbes–Wallis dispute, *the Royal Society journal of the history of science* (Notes and Records) **72** (2018), 469–486.

147. D. Kahneman and A. Tversky, Prospect Theory: An Analysis of Decision under Risk, *Econometrica*, Vol. 47, No. 2 (Mar., 1979), 263–292.

148. E. Kalai and M. Smorodinsky, Other solutions to Nash's bargaining problem, *Econometrica* **43** (1975), 513–518.

149. J. Kapeller and S. Steinerberger, How Formalism shapes Perception: An Experiment on Mathematics as a Language, *International Journal of Pluralism and Economics Education* **4** (2013): 138–156.

150. J. Kapeller and S. Steinerberger, Modeling the evolution of preferences: an answer to Schubert and Cordes, *Journal of Institutional Economics* **10** (2014), 337–347.

151. K. Katz and M. Katz, Cauchy's continuum, *Perspectives on science* **19**(4), 426–452.

152. A. Kawamura, A. Moriyama, Y. Otachi and J. Pach, A lower bound on opaque sets, *Computational Geometry* **80** (2019), 13–22.

153. J. Kepler, *Tabulae Rudolphinae, quibus astronomicæ scientiae, temporum longinquitate collapsæ restauratio continetur; a phoenice illo astronomorum tychone* [...], Jonae Saurii, 1627, Retrieved from the Library of Congress, item 49038330.

154. W. Kessen, Rousseau's Children, *Daedalus* **107** (1978), 155–166.

155. R. Kim, The Mind-Body Nexus in Beckett's Writing of Love-Melancholy, *Samuel Beckett Today/Aujourd'hui* **29** (2017), 37–50.

156. F. Klein, *Development of mathematics in the 19th century*, Math Sci Press, 1979,

157. K. Kohli, Aus dem Briefwechsel zwischen Leibniz und Jakob Bernoulli, In: *Die Werke von Jakob Bernoulli. Vol. 3*, Birkhäuser Basel 1975, 509–513.

158. H. Kornhuber and L. Deecke, Hirnpotentialänderungen bei Willkürbewegungen und passiven Bewegungen des Menschen: Bereitschaftspotential und reafferente Potentiale, *Pflüger's Archiv für die gesamte Physiologie des Menschen und der Tiere* **284** (1965): 1–17.

159. G. Kreisel, Kurt Gödel, 28 April 1906 - 14 January 1978, *Biographical Memoirs of Fellows of the Royal Society* **26**, 1980.

160. D. Laertius, *Lives and Opinions of Eminent Philosophers*, translated by: C.D. Yonge. London: Henry G. Bohn, 1853.

161. J. C. Lagarias, *The Ultimate Challenge: The $3x+1$ Problem*, American Mathematical Society, 2023.

162. J. L. Lagrange, Demonstration d'un theoreme nouveau concernant les nombres premiers, *Nouveaux Mémoires de l'Académie Royale des Sciences et Belles-Lettres (Berlin)*, vol. 2, pages 125–137 (1771).

163. P. S. Laplace, *Essai philosophique sur les probabilites*, Paris: Courcier, 1814.

164. P. Lax, email to Scott Stankey, Oct 28 2017, personal communication.

165. M. Lazzarini, Un'applicazione del calcolo della probabilita alla ricerca sperimentale di un valore approssimato di π, *Periodico di Matematica per l'Insegnamento Secondario* **4** (1901), 140–143.

166. A.-M. Legendre, *Nouvelles methodes pour la determination des orbites des cometes* [*New Methods for the Determination of the Orbits of Comets*] (in French), Paris: F. Didot, 1805.

167. Gottfried Wilhelm Freiherr von Leibniz, *Opera Philosophica Quae Exstant Latina, Gallica Germanica Omnia*, vol 1, Eichler, 1840.

168. G. W. Leibniz, Reply to the comments in the second edition of M. Bayle's Critical Dictionary, in the article 'Rorarius', concerning the system of pre-established harmony, in G. W. Leibniz: *Philosophical texts* (1998) (R. Francs & R. S. Woolhouse, Trans.), pp. 241–253. Oxford University Press.

169. *Briefwechsel zwischen Leibniz und Christian Wolf; aus den Handschriften der Koeniglichen Bibliothek zu Hannover herausgegeben von Gerhardt*, Schmidt, Halle, 1860.

170. C. G. Lekkerkerker, *Voorstelling van natuurlijke getallen door een som van getallen van Fibonacci*, Stichting Mathematisch Centrum. Zuivere Wiskunde, (ZW 30/51), 1951.

171. J. Leurechon, *Selectae propositiones in tota sparsim mathematica pulcherrimae. Ad usum et exercitationem celebrium academiarum*, G. Bernardus, 1629.

172. M. Lewis, *The undoing project: A friendship that changed the world*, Penguin UK, 2016.

173. Obituary: Edward Hubert Linfoot, *Bulletin of the London Mathematical Society* **16** (1984), 52–58.

174. R. J. Lipton and K. W. Regan, Roger Apéry: Explaining Proofs, in *People, Problems, and Proofs: Essays from Gödel's Lost Letter*, 2010, pp. 281–285.

175. W. Lloyd, *Two Lectures on the Checks to Population*, Oxford, 1833

176. E. Loomis, *The Pythagorean Proposition: Its Proofs Analyzed and Classified and Bibliography of Sources for Data of the Four Kinds of Proofs*, Mohler Printing Company, 1927.

177. E. Lorenz, Predictability: does the flap of a butterfly's wings in Brazil set off a tornado in Texas? American Association for the Advancement of Science. In 139th meeting, vol. 29, 1972.

178. Edward N. Lorenz, *The Essence of Chaos*, University of Washington Press, 1995

179. M. Losada, The complex dynamics of high performance teams, *Mathematical and computer modelling* **30** (1999), 179–192.

180. R. Luce and H. Raiffa, *Games and decisions*, John Wiley and Sons, New York, 1957.

181. Lucretius, *On the Nature of Things*, translated by William Ellery Leonard. E. P. Dutton, 1916.

182. J. Lützen, Sturm and Liouville's work on ordinary linear differential equations. The emergence of Sturm-Liouville theory, *Archive for history of exact sciences* **29** (1984), 309–376.

183. N. Machiavelli, *The Prince*, translated by J. Scott Byerley, Sherwood, Neely and Jones, London, 1810.

184. N. Macrae, *John von Neumann: The Scientific Genius Who Pioneered the Modern Computer, Game Theory, Nuclear Deterrence, and Much More*, Plunkett Lake Press, 2016.

185. MacTutor Biography project, created and maintained by Edmund Robertson and John O'Connor of the School of Mathematics and Statistics at the University of St Andrews.

186. P. Mancosu and E. Vailati, Torricelli's infinitely long solid and its philosophical reception in the seventeenth century, *Isis* **82** (1991), 50–70.

187. B. Mandelbrot, The Variation of Certain Speculative Prices, *The Journal of Business*, Vol. 36, No. 4 (Oct., 1963), 394–419

188. B. Mandelbrot, *The fractalist: Memoir of a scientific maverick*, Vintage, 2012.

189. Pierre Louis Moreau de Maupertuis, Les Loix du mouvement et du repos déduites d'un principe metaphysique, *Histoire de l'Académie Royale des Sciences et des Belles Lettres* (1746), pp. 267–294.

190. Francesco Maurolico, *Arithmeticorum libri duo*, Apud Franciscum Franciscium Senesem, Venetia, 1575.

191. B. McGinn, The Conjecture on the Last Days (Coniectura de ultimis diebus): Cusanus and Concordist Eschatology, In: *Nicholas of Cusa and Times of Transition*, Studies in the History of Christian Traditions, Volume 188, 2019, pp. 334–345.

192. K. McNamee and M. Jacovides, Annotations to the Speech of the Muses (Plato "Republic" 546B-C), *Zeitschrift für Papyrologie und Epigraphik* (2003), 31–50.

193. H. Meinhardt, *The Algorithmic Beauty of Sea Shells*, Springer, 2009.

194. P. Mengoli, *Novae quadraturae arithmeticae, seu De additione fractionum*, Bononiae, ex Typographia Iacobi Montij, 1650.

195. N. Metropolis, The beginning of the Monte Carlo method, *Los Alamos Science* **15** (1987), 125–130.

196. J. S. Mill, *Autobiography*, Columbia University Press, 1924.

197. J. S. Mill, *Utilitarianism*, reprinted from 'Fraser's Magazine' (**64**, Issue 382, 1861), 7th edition, Longmans, Green and Co, London, 1879.

198. F. C. Mills, *The Behavior of Prices*, National Bureau of Economic Research, 1927.

199. W. C. Mitchell, *The making and using of index numbers*, Washington DC: US Government Printing Office, 1938.

200. G. Mittag-Leffler, The Higher Mathematics, *Nature*, Volume 32 (1885), 302–303.

201. G. Monge, Mémoire sur la théorie des déblais et des remblais, *Histoire de l'Académie Royale des Sciences de Paris, avec les Mémoires de Mathématique et de Physique pour la meme anné*, pages 666–704, 1781.

202. R. Monk, *Ludwig Wittgenstein: The Duty of Genius*, Vintage, 1991.

203. Pierre Rémond de Montmort, *Essay d'analyse sur les jeux de hazard*, J. Quillau (Paris), 1713.

204. G. E. Moore, *Philosophical Papers*, London Allen & Unwin, 1959.

205. Rosemarie Nagel, Unraveling in Guessing Games: An Experimental Study, *The American Economic Review*, Vol. 85, No. 5 (Dec., 1995), 1313–1326.

206. J. Napier, *A plaine discouery of the whole Reuelation of Saint Iohn set downe in two treatises: the one searching and prouing the true interpretation thereof: the other applying the same paraphrastically and historically to the text. Set foorth by Iohn Napeir L. of Marchistoun younger. Whereunto are annexed certaine oracles of Sibylla, agreeing with the Reuelation and other places of Scripture*, Printed by Robert Walde-grane, Edinburgh, 1593, scan by Internet Archive, ark:/13960/t03x9gn74.

207. *Mémoires pour servir à l'Histoire de France sous Napoléon, écrits a Sainte-Hélìe, Tome premier, écrit par le Général Gourgaud, son aide-de-camp*, Paris, 1823.

208. S. Nasar, *A Beautiful Mind*, Simon & Schuster, 1998.

209. John Nash, The Bargaining Problem, *Econometrica* **18** (1950), 155–162.

210. R. Nelson, *Proofs without words: Exercises in visual thinking*, Mathematical Association of America, 1993.

211. J. v. Neumann, Zur Theorie der Gesellschaftsspiele, *Mathematische Annalen* **100**(1) (1928), 295–320.

212. J. von Neumann, *Mathematical Foundations of Quantum Mechanics*, Springer, 1932.

213. J. von Neumann and O. Morgenstern, *Theory of Games and Economic Behavior*, Princeton University Press, 1944.

214. E. Newbold, Practical Applications of the Statistics of Repeated Events, Particularly to Industrial Accidents, *Journal of the Royal Statistical Society*, Vol. 90, No. 3 (1927), 487–547.

215. Nicomachi Geraseni Pythagorei, *Introductionis arithmeticae Libri II*, Teubner, 1866.

216. M. Olivier, *Les nombres indices de la variation des prix*, M. Giard, 1927.

217. Nicole Oresme, *Le Livre du ciel et du monde*, Edited by Albert D. Menut and Alexander J. Denomy, The University of Wisconsin Press; First Edition (January 1, 1968).

218. A. Ottolini and S. Steinerberger, Greedy Matching in Optimal Transport with concave cost, arXiv:2307.03140.

219. W. Paley, *Natural theology: or, Evidences of the existence and attributes of the Deity, collected from the appearances of nature*, Sheldon, New York, 1879.

220. *The Commentary of Pappus on Book X of Euclid's Elements*, Arabic Text and Translation by William Thomson with Introductory Remarks, Notes and a Glossary of Technical Terms by Gustav Junge and William Thomson, Harvard University Press, 1930

221. *Matthaei Paris, monachi Albanensis Angli, Historia major: juxta exemplar Londinense 1571 verbatim recusa. Et cum Rogeri Wendoveri, Willielmi Rishangeri, authorisque majori minorique historiis chronicisque mss. in Bibliotheca Regia, Collegii Corporis Christi Cantabrigiae, Cottoniaque fideliter collata : huic primum editioni accesserunt, duorum Offarum Merciorum regum, & viginti trium Abbatum S. Albani vitae : una cum libro Additamentorum per eundem authorem, 1640; Londini : Excudebat Richardus Hodgkinson : Sumptibus Cornelii Bee & Laurentii Sadler, in vico vulgariter dicto Little Britaine.*

222. J. Parton, *Life and Times of Benjamin Franklin* (Vol. 52), Mason Brothers, 1864.

223. B. Pascal, *Pensees*, A Dutton Paperback, New York, 1958.

224. P. Pasles, The Lost Squares of Dr. Franklin: Ben Franklin's Missing Squares and the Secret of the Magic Circle, *The American Mathematical Monthly* **108**(6) (2001), 489–511.

225. G. Peano, Sur une courbe, qui remplit toute une aire plane, *Mathematische Annalen* **36** (1890), 157–160.

226. G. Peano, De Latino Sine Flexione. Lingua Auxiliare Internationale, *Revista de Mathematica (Revue de Mathématiques)*, Tomo VIII, Fratres Bocca Editores: Torino, 1903, 74–83.
227. K. Pearson, Laplace, *Biometrika* **21**(1-4) (1929), 202–216.
228. *Plato in Twelve Volumes*, Vol. 12 translated by Harold N. Fowler. Cambridge, MA, Harvard University Press; London, William Heinemann Ltd. 1921.
229. *Plato: The Collected Dialogues*, Translation by Paul Shorey, Eds. Edith Hamilton & Huntington Cairns, Princeton University Press, Princeton, N. J., 1961.
230. *The Theaetetus of Plato: A translation with an introduction.* Dyde, S. W. (Ed.), James Maclehose and Sons, 1899.
231. Plato, *The Republic*, translated by B. Jowett, Oxford University Press, 1888.
232. Plato, *Laws*, translated by B. Jowett, Provided by The Internet Classics Archive.
233. *Plautus, with an English translation by Paul Nixon*, Harvard University Press, 1916.
234. K. Plofker, Mathematics in India, In Katz, Victor J (ed.), *The Mathematics of Egypt, Mesopotamia, China, India, and Islam: A Sourcebook*, Princeton University Press, 2007.
235. Henri Poincaré, Sur la dynamique de l'electron. Note de H. Poincaré. *C.R.* **140** (1905), 1504–1508.
236. W. Poundstone, *Prisoner's dilemma*, Anchor, 2011.
237. S. Probst, Infinity and creation: the origin of the controversy between Thomas Hobbes and the Savilian professors Seth Ward and John Wallis, *British J. Hist. Sci.* **26** (90, 3) (1993), 271–279.
238. A. Quetelet, *On Man and the Development of his Faculties, or Essay on Social Physics*, W. & R. Chambers, 1842.
239. S. Ramanujan, Notebook 1, Chapter 8, Page 3, Tata Institute of Fundamental Research (TIFR), Mumbai, 1957.
240. J. Rawls, *A Theory of Justice: Revised Version*, Harvard University Press, 1999.
241. C. Reid, *Hilbert*, Springer Science & Business Media, 1996.
242. B. Rittaud and A. Heeffer, The pigeonhole principle, two centuries before Dirichlet, *The Mathematical Intelligencer* **36** (2014), 27–29.
243. S. Roberts, *Genius at Play: The Curious Mind of John Horton Conway*, Bloomsbury USA, 2015.
244. S. Roberts, The Lasting Lessons of John Conway's Game of Life, *New York Times* 12/28/2020.
245. P. Rosenbloom, *The Elements of Mathematical Logic*, New York: Dover Publications, 1950.
246. Gian-Carlo Rota, *Indiscrete Thoughts*, Springer, 1997.
247. *The Confessions of Jean Jacques Rousseau, now for the first time completely translated into English without expurgation*, Privately Printed, 1904.
248. P. Ruju and M. Mostert, *The life and times of Guglielmo Libri (1802–1869): Scientist, patriot, scholar, journalist, and thief: a nineteenth-century story*, Verloren Publishers, 1995.
249. B. Russell, Mathematical Logic as Based on the Theory of Types, *American Journal of Mathematics* **30** (1908), 222–262.
250. B. Russell, Ludwig Wittgenstein, *Mind* **60** (239), 1951.
251. V. Salikhov, On the Irrationality Measure of pi, *Usp. Mat. Nauk* **63**, 163–164, 2008. English transl. in *Russ. Math. Surv.* **63**, 570–572, 2008.
252. Ferdinand de Saussure, *Cours de linguistique générale*, Payot, Paris, 1916.
253. D. Schattschneider, The mathematical side of MC Escher, *Notices of the AMS* **57** (2010), 706–718.
254. Dierk Schleicher, Interview with John Horton Conway, *Notices of the AMS* **60**, May 2013.
255. A. Schoenflies, Die Krisis in Cantor's mathematischem Schaffen, *Acta Mathematica* **50** (1927), 1–23.
256. A. Schopenhauer, *Preisschrift ueber die Freiheit des Willens*, 1978.
257. J. F. Scott, The Reverend John Wallis, F.R.S. (1616–1703), *Notes and Records of the Royal Society of London*, Vol. 15 (Jul., 1960), 57–67.
258. Johannes Duns Scotus, *In VIII. libros Physicorum Aristotelis Quaestiones*, Apud Ioannem Crithium, sub signo Galli, Colonia, 1618, Bayerische Staatsbibliothek München, BV001484785, urn:nbn:de:bvb:12-bsb10215618-7.

259. E. Seneta, Mathematics, religion, and Marxism in the Soviet Union in the 1930s, *Historia Mathematica* **31**, Issue 3, August 2004, Pages 337–367.

260. E. Seneta, Statistical Regularity and Free Will: L.A.J. Quetelet and P.A. Nekrasov, *International Statistical Review* **71** (2007), 319—334.

261. Sextus Empiricus, *Outlines of Pyrrhonism*, trans. Robert Gregg Bury (Loeb ed.) (London: W. Heinemann, 1933).

262. L. Sigler, *Fibonacci's Liber Abaci: A Translation into Modern English of Leonardo Pisano's Book of Calculation*, Sources and Studies in the History of Mathematics and Physical Sciences, Springer, 2002.

263. P. Singer and S. Chao-Hwei, *The Buddhist and the Ethicist: Conversations on Effective Altruism, Engaged Buddhism, and How to Build a Better World*, Shambhala, 2023.

264. D. Singmaster, How Often Does an Integer Occur as a Binomial Coefficient?, *The American Mathematical Monthly*, Vol. 78, No. 4 (Apr., 1971), 385–386.

265. David Singmaster, Interview, Gathering 4 Gardner 13, interview conducted April 13, 2018, available on YouTube.

266. J. Soni and R. Goodman, *A Mind at Play: How Claude Shannon Invented the Information Age*, Simon & Schuster, 2017.

267. S. Steinerberger, A Hidden Signal in the Ulam sequence, *Experimental Mathematics* **23**, 460–467 (2017).

268. S. Steinerberger, An amusing sequence of functions, *Mathematics Magazine* **91** (2018), 262–266.

269. S. Steinerberger, Quadratic Crofton and sets that see themselves as little as possible, *Monatshefte für Mathematik* **204**(2) (2024), 1–13.

270. S. Steinerberger, Random Growth via Gradient Flow Aggregation, *Journal of Applied Probability*, to appear.

271. S. Steinerberger and T. Zeng, A curious dynamical system in the plane, arXiv:2409.08961.

272. Michael Stifel, *Ein Rechen Buchlin vom Endchrist. Apocalypsis In Apocalypsim*, Georg Rhaw, Wittenberg, 1532.

273. Michael Stifel, *Arithmetica Integra*, apud Iohan Petreium, Norimbergae, 1544, Courtesy of The Linda Hall Library of Science, Engineering & Technology.

274. S. Stigler, Gieryn, F. (ed.), Stigler's law of eponymy, *Transactions of the New York Academy of Sciences* **39** (1980), 147–58.

275. M. Van Strien, On the origins and foundations of Laplacian determinism, *Studies in History and Philosophy of Science Part A* **45** (2014), 24–31.

276. O. Taussky-Todd, Remembrances of Kurt Gödel, in: *Gödel Remembered*, Napoli, Bibliopolis, 1987.

277. J. Thomson, Tasks and Super-Tasks, *Analysis* **15**(1) (1954), 1–13.

278. Thorvald Thiele, *Sur la compensation de quelques erreurs quasi-systématiques par la méthode des moindres carrés*, Copenhague, C. A. Reitzel, 1880.

279. V. M. Tikhomirov, The life and work of Andrei Nikolaevich Kolmogorov, *Russian Mathematical Surveys* **43**(6) (1988), 3–33.

280. E. Torricelli, *Lezioni accademiche* (Vol. 139), G. Silvestri, 1823.

281. E. Toulouse, *Henri Poincare par le Dr. Toulouse*, Flammarion, 1910.

282. A. Turing, Computing Machinery and Intelligence, *Mind* **49** (1950), 433–460.

283. A. Tversky and D. Kahneman, Rational Choice and the Framing of Decisions, *The Journal of Business*, Vol. 59, No. 4, Part 2: The Behavioral Foundations of Economic Theory (Oct., 1986), pp. S251–S278.

284. C. Tweedie, *James Stirling; a sketch of his life and works along with his scientific correspondence*, Clarendon Press, Oxford, 1922.

285. John Tzetzes, *Historiarum variarum chiliades Graece*, edited by Theophilus Kiesslingius, Lips, 1826, translated by A. Untila, G. Berkowitz, K. Ramiotis, V. Dogani, J. Alexander, M. Fadhlurrahman, N. Giallousis.

286. S. Ulam, *Adventures of a Mathematician*, University of California Press, 1991.

287. S. Ulam, Combinatorial Analysis in Infinite Sets and Some Physical Theories, *SIAM Review* **6**(4) (1964), 343–355.

288. G. Vacca, Maurolycus, the first discoverer of the principle of mathematical induction, *Bull. Amer. Math. Soc.* **16** (1909): 70–73.

289. J. Vandehey, On the binary digits of $\sqrt{2}$, arXiv:1711.01722.

290. M. G. Vann, Of rats, rice, and race: The great Hanoi rat massacre, an episode in French colonial history, *French Colonial History* **4** (2003), 191–203.

291. Frans de Waal, *Primates and Philosophers: How Morality Evolved*, Princeton University Press, 2009.

292. H. Wang, Some Facts about Kurt Gödel, *The Journal of Symbolic Logic* **46** (1981), 653–659.

293. Dr. John Wallis his answer, by way of letter to the publisher, to the book, entituled Lux Mathematica, &c. described in Numb. 86. of these tracts, *Philosophical Transactions of the Royal Society of London*, Volume 7, Issue 87, Oct 1672.

294. E. Waring, *Meditationes Algebraicae*, Cantabridgiae, 1770.

295. A. Waugh, *The house of Wittgenstein: A family at war*, A&C Black, 2009.

296. H. Weaver, Object lessons: a cultural genealogy of the dunce cap and the apple as visual tropes of American education, *Paedagogica Historica: International Journal of the History of Education* **48** (2012), 215–241.

297. J. Weizenbaum, ELIZA–A Computer Program for the Study of Natural Language Communication Between Man and Machine, *Communications of the ACM* **9** (1966), 36–45.

298. J. Weizenbaum, *Computer Power and Human Reason: From Judgment to Calculation*, W. H. Freeman and Company, 1976.

299. E. Wigner, The unreasonable effectiveness of mathematics in the natural sciences, In *Mathematics and science* (pp. 291–306), 1990.

300. L. Wittgenstein: *Tractatus Logico-Philosophicus*, First published by Kegan Paul (London), 1922.

301. L. Wittgenstein: *The Big Typescript* (TS 213), Edited by Luckhardt and Aue, Wiley 2012.

302. L. Wittgenstein, *Ueber Gewissheit*, Suhrkamp, 1970.

303. L. Wittgenstein, *Wittgenstein's Lectures on the Foundations of Mathematics, Cambridge, 1939*: From the Notes of RG Bosanquet, Norman Malcolm, Rush Rhees, and Yorick Smythies. Diamond, C. (Ed.), Cornell University Press, 1976.

304. S. Wolfram, Statistical mechanics of cellular automata, *Rev. Mod. Phys.* **55** (1983), 601–644.

305. Andrew Young, Did Monge really explain inferior mirages?, *Comptes Rendus. Physique* **23**(S1) (2022), 1–15.

306. A. N. Zaydel, Delusion or fraud, *Quantum*, September/October, 1990, 6–9.

307. E. Zeckendorf, Representation des nombres naturels par une somme de nombres de Fibonacci ou de nombres de Lucas, *Bull. Soc. Roy. Sci. Liege* **41** (1972), 179–182.

308. E. C. Zeeman, On the dunce hat, *Topology* **2** (1963), 341–358.

309. Y. B. Zel'Dovich, On the theory of unstable states, *Sov. Phys. JETP* **12** (1961), 542–545.

310. E. Zermelo, Hobson, E. W.; Love, A. E. H. (eds.), Uber eine Anwendung der Mengenlehre auf die Theorie des Schachspiels [On an Application of Set Theory to the Theory of the Game of Chess], *Proceedings of the Fifth International Congress of Mathematicians* (1912), Cambridge: Cambridge University Press, pp. 501–504.

Index

MIX
Papier aus verantwortungsvollen Quellen
Paper from responsible sources
FSC® C105338

If you have any concerns about our products,
you can contact us on
ProductSafety@springernature.com

In case Publisher is established outside the EU,
the EU authorized representative is:
**Springer Nature Customer Service Center GmbH
Europaplatz 3, 69115 Heidelberg, Germany**

Printed by Libri Plureos GmbH
in Hamburg, Germany